UNIVERSITY OF STRATHCLYDE

30125 00531506 3

KV-038-889

Books are to be returned on or before
the last date below.

Toxic Impacts of Wastes on the Aquatic Environment

Toxic Impacts of Wastes on the Aquatic Environment

Edited by

J. F. Tapp
ZENECA, Brixham Environmental Laboratory, UK

J. R. Wharfe
Environment Agency, Worthing, UK

S. M. Hunt
Hyder Environmental, Runcorn, UK

The Proceedings of an International Conference sponsored by the Water Chemistry Forum of The Royal Society of Chemistry, the Chartered Institution of Water and Environmental Management, Zeneca Brixham Environmental Laboratory, NRA/HMIP/SNIFFER, the Centre for Hazard and Risk Management at Loughborough University of Technology, and the Research Centre at the University of Luton on Toxic Impacts of Wastes on the Aquatic Environment. The meeting was held on 14–17 April 1996 at Loughborough University of Technology.

D
628.168
TOX

Special Publication No. 193

ISBN 0-85404-781-6

A catalogue record for this book is available from the British Library

© The Royal Society of Chemistry 1996

All rights reserved.
Apart from any fair dealing for the purposes of research or private study, or criticism or review as permitted under the terms of the UK Copyright, Designs and Patents Act, 1988, this publication may not be reproduced, stored or transmitted, in any form or by any means, without the prior permission in writing of The Royal Society of Chemistry, or in the case of reprographic reproduction only in accordance with the terms of the licences issued by the Copyright Licensing Agency in the UK, or in accordance with the terms of the licences issued by the appropriate Reproduction Rights Organization outside the UK. Enquiries concerning reproduction outside the terms stated here should be sent to The Royal Society of Chemistry at the address printed on this page.

Published by The Royal Society of Chemistry,
Thomas Graham House, Science Park, Milton Road, Cambridge CB4 4WF, UK

Typeset by Computape (Pickering) Limited, North Yorkshire
Printed by Hartnolls Ltd, Bodmin, Cornwall, UK

Preface

People are probably more aware of environmental issues now than they have been at any other time in the history of mankind. It is reasonable to say that, given currently available treatment technologies, the production of waste matter containing potentially toxic contaminants, and its subsequent release to the environment, is inevitable in a consumer society. Regulatory procedures require the elimination of particularly hazardous substances and the stringent control of others which are released to the environment. The toxic impact of wastes to the aquatic environment was the subject of these proceedings of the conference held at Loughborough University in April 1996. Platform and poster presentations by authors from the United Kingdom, Europe and the United States contributed to this internationally important meeting.

Contaminants enter the aquatic environment from many different sources. Some, such as point source discharges, are more easy to control than others which arise from diffuse inputs. To a large extent, the control of effluent discharges to the aquatic environment has traditionally been based on chemical limits. There are many thousands of potentially harmful substances, of which only a very few have standards which have been derived to protect aquatic life. This fact, and also because it is not possible to predict additive and synergistic toxic effects of mixtures often found in complex discharges, has increased efforts to develop and apply more meaningful biological and ecotoxicological measures to assess their impact when released to the environment.

During the course of the conference many aspects concerned with the toxic impact of wastes on the aquatic environment were considered, these included: current legislation in the United Kingdom and Europe; toxic measures of environmental damage and the development of control procedures; new and novel methods to improve and provide more rapid assessment of toxicity; the application of quality control procedures to ensure excellent information from test procedures; and a number of case studies demonstrating environmental damage arising from toxic inputs and others on toxicity identification and reduction evaluation programmes. Additionally, political and economic implications were considered with a view to the further development of methods and procedures to help improve and protect aquatic life.

These proceedings present current information on regulatory controls and impact assessments of toxic waste entering the aquatic environment and represent a step forward in our understanding of these scientific and technically complex environmental issues. The success of the conference was due mainly to the hard work of the organising committee and the high standard of the presentations. The editors express their gratitude to the sponsors of the conference and their sincere thanks to all who have contributed to these proceedings.

The organising committee comprised: Chairman, Dr Colin Fuller (Loughborough University); and members; Don Bealing (Hazelton Europe), Dr Paul Beckwith (British Waterways), Dr Keith Colquhoun (Thames Water), David Fearnside (Yorkshire Water), Lavinia Gittins (CIWEM), Dr Peter Hiley (Yorkshire Water), Stephen Hunt (Hyder Environmental (previously, Acer Environmental)), Professor David Rawson (University of Luton), Dr Richard Sharp (CAMR, Porton Down), John Tapp (Zeneca Brixham Environmental Laboratory), Dr Clive Thompson (LabServices, Yorkshire Environmental), Dr James Wharfe (Environment Agency), and Dr Paul Whitehouse (WRC).

The conference was sponsored by: the Water Chemistry Forum of The Royal Society of Chemistry; The Chartered Institution of Water and Environmental Management; Zeneca Brixham Environmental Laboratory; the Environment Agency; the Scottish and Northern Ireland Forum for Environmental Research; the Centre for Hazard and Risk Management at Loughborough University; and The Research Centre at the University of Luton. The International Association of Water Quality also supported the meeting.

John F Tapp
James R Wharfe
Stephen M Hunt

Contents

Preface
J. F. Tapp, J. R. Wharfe and S. M. Hunt v

The Toxic Impacts of Wastes from Source to Ultimate Fate
Sir Hugh Rossi 1

The Scientific, Economic, Environmental and Political Significance of the Discharge of Toxic Wastes to the Aquatic Environment
D. J. H. Phillips 9

Toxicity Based Criteria for the Regulatory Control of Waste Discharges and for Environmental Monitoring and Assessment in the United Kingdom
J. R. Wharfe 26

The Use of Direct Toxicity Assessment to Control Discharges to the Aquatic Environment in the United Kingdom
D. Tinsley, I. Johnson, R. Boumphrey, D. Forrow and J. R. Wharfe 36

The Precision of Aquatic Toxicity Tests: Its Implications for the Control of Effluents by Direct Toxicity Assessment
P. Whitehouse, P. A. H. van Dijk, P. J. Delaney, B. D. Roddie, C. J. Redshaw and C. Turner 44

Amtox^{TM} - A New Concept for Rapid Nitrification Inhibition Testing Applicable to the Laboratory and On-line at Treatment Works
J. Upton and S. R. Pickin 54

Performance Trials of a New Automated Respirometer (MERIT 20) in Performing Respiration and Nitrification Inhibition Analyses
M. R. Bolton, D. Fearnside and R. A. Addington 64

The Potential for Biosensors to Assess the Toxicity of Industrial Effluents
J. G. Rogerson, A. Atkinson, M. R. Evans and D. M. Rawson 74

Application of Routine Ecotoxicological Screening Methods for Assessing Suitability of Trade Effluents for Biological Treatment
K. Wadhia, A. Colley and K. C. Thompson   84

Eclox™: A Rapid Screening Toxicity Test
E. Hayes and M. Smith   94

Interpretation, Relevance and Extrapolations: Can We Devise Better Ecotoxicological Tools to Assess Toxic Impacts?
M. H. Depledge   104

In situ Assays for Monitoring the Toxic Impacts of Waste in Rivers
M. Crane, I. Johnson and L. Maltby   116

Sub-lethal Biological Effects Monitoring Using the Common Mussel (*Mytilus edulis*): Comparison of Laboratory and *In situ* Effects of an Industrial Effluent Discharge
B. D. Roddie, C. J. Redshaw and S. Nixon   125

Sub-lethal Effects of Waste on Marine Organisms: Responses Measured at the Whole Organism Level
P. Donkin, J. Widdows, D. M. Lowe, M. E. Donkin and D. N. Price   138

The Toxic Impact of Petro-chemical and Oil-refining Waste on Hydrobionts
N. G. Kuramshina, E. M. Kuramshin and S. V. Pavlov   146

The Role of Environmental Quality Standards in Controlling Chemical Contaminants in the Environment
N. G. Cartwright and S. Lewis   149

When the Solution is More Toxic than the Pollution
J. H. Churchley and E. M. Hayes   157

Possible Ways of Detoxification of Heavy Metals (Zinc, Lead and Cadmium) Discharged with Wastes in High-Bioproductive Reservoirs
I. V. Iskra and P. N. Linnik   167

United States Environmental Protection Agency's Water-quality Based Approach to Toxics Control
M. A. Heber and T. J. Norberg-King. Editorial synopsis by J. F. Tapp, J. R. Wharfe and S. M. Hunt   175

The Ecotoxicological Assessment of Hydrocarbons in an Urban Aquatic System
R. H. Jones, D. M. Revitt, R. B. E. Shutes and J. B. Ellis   188

Contents ix

Selection of Toxicity Assays for Multiple Test Protocols Used in
Wastewater Quality Control and Treatment Optimisation
P. Hiley 200

The Use of Toxicity Identification Evaluation on Industrial Effluent
Discharges
V. T. Coombe, K. W. Moore and M. J. Hutchings 208

Overall Cost Savings and Treatment Optimisation by Segregation of the
Component Waste Streams in a Toxicity Reduction Evaluation
J. H. H. Looney, M. A. Collins, M. S. Stein and S. R. Harper 217

Toxicity Testing as a Practical Tool for Monitoring Industrial Effluents –
a Case Study
D. C. Watson, P. Fawcett and N. Gudgeon 230

Toxicity Reduction Measures for a Synthetic Fibre Industry Effluent –
a Case Study
J. L. Musterman, T.H. Flippin and G. W. Pulliam 238

What Next? – Future Developments in the Field of Aquatic
Environmental Toxicology
I. Johnson 251

Posters

Changes in Toxicity of Ground and Surface Waters during Remediation
of a Former Gasworks Site
A. J. Hart and A. Trim 260

Rapid Determination of Heavy Metal Toxicity in Sewage Sludge Using a
Bioluminescence-Based Bioassay
*S. Sousa, C. D. Duffy, E. A. S. Rattray, K. Killham, L. A. Glover,
D. Fearnside, K. C. Thompson and M. S. Cresser* 261

Fate and Environmental Effect of Produced Water Discharged from the
Clyde Oil Production Platform
I. Vance, E. J. Butler and S. A. Flynn 263

Remediation of Contaminated Dredged Sediments by Physical Processing
S. T. Hall, A. Burton and B. Denby 264

Toxicity Assessment of Complex Industrial Effluents Discharged
to Sewer
M. Arretxe, G. Ellen, P. Tetlaw, M. Heap and N. Cristofi 265

New Assays for Inhibition of Nitrification and Denitrification – A
Comparative Study Applied to Industrial Wastwater
C. Grunditz, L. Gumaelius and G. Dalhammar 267

Ecotoxicological and Chemical Measurements of Municipal Wastewater
Treatment Plant Effluents
J. Garric, B. Vollat, D. K. Nguyen, M. Bray, B. Migeon and A. Kosmala 269

Fish EROD Activity to Assess the Impact of a Wastewater Treatment
Plant. Field and On-site Experimental Approaches
A. Kosmala, B. Migeon and J. Garric 271

Influence of Organic Matter on the Response of Microtox Tests to Three
Toxic Compounds
C. Ravelet, B. Vollat, J. Garric and B. Montuelle 273

The Relationship between Measured Concentrations of Contaminants and
the Toxicity of Leachates from a Contaminated Landfill Site
M. J. Mallett and J. Sweeney 274

Stonewort Cell: A Biosensor to Assess Wastewater Toxicity of Membrane
and Whole Cell Levels
L. Manusadzianas and R. Vitkus 276

Development of Rapid Toxicity Assay for Paper *via* Microtox™ Test
A. Kahru, L. Pollumaa, I. Kulm and E. Paart 278

Optical Fibre Toxicity Bio-sensor
D. F. Merchant, P. J. Scully, R. Edward and J. Grabowski 279

The Influence of Sewage Treatment Processes on the Oestrogenicity of
Chemicals to Fish
G. Panter, R. Thompson and J. Sumpter 280

Evaluation of Two Enhanced Chemiluminescene Test Kits for Water
Quality and Toxicity Testing
A. Colley, K. Wadhia and K. C. Thompson 282

The Fungicide Busan 30WB Causes Lipid Peroxidation in Mussels
Exposed to Leather Tannery Effluent
A. R. Walsh, P. Byrne and J. O'Halloran 284

Monitoring Compliance with a Toxicity-based Consent – The Case for
Limit Tests
I. Johnson and P. Whitehouse 285

The Application of QSARs to Predict Environmental Hazard of
Chemicals
A. B. A. Boxall 286

Ecogeochemical Consequences of Contamination of the Nura River
in Central Kazakhstan by Mercury-containing Wastewater from
Acetaldehyde Production
M. Ilyuschchenko, S. Butlatkulov and S. Heaven 287

Subject Index 289

The Toxic Impacts of Wastes from Source to Ultimate Fate

Sir Hugh Rossi

Simmons and Simmons, 21 Wilson Street, London, EC2M 2TX, UK

ABSTRACT

The conference theme is all aspects of toxic wastes polluting surface waters, groundwaters and the sea. Much of the substance of this conference will be about the tests and testing philosophies associated with the control of these wastes, throughout their life cycle. Wastes may arise from industrial processes discharging via municipal treatment, leachates from waste tips, aerial deposition and probably many other ways also. Detecting the degree of toxicity in a sufficiently precise way relevant to the receiving watercourse or sea area has been a problem which is perhaps now overcome, at least sufficiently to allow the enforcement of toxicity based consent standards in several countries.

The development of control legislation for various types of waste will be discussed, setting the background to today's situation. The division of the former Water Authorities into regulatory and operational parts came from a need to have firmer control of river pollution. Since its formation the National Rivers Authority has had only partial control, and this year the Environment Agency succeeds the NRA, drawing together various bodies involved in the control of toxic wastes. The increasing influence of European legislation is reflected in a discussion of the differences in concept between English and European law regarding these matters. The future legislation, ever tightening, will be discussed in relation to the present situation.

1 INTRODUCTION

Toxic substances, and the routes by which they may reach and contaminate the aquatic environment, are several and varied. They may be gaseous in origin, generated by the burning of fossil fuels, whose acidity can have fatal consequences for life in the rivers and lakes into which their depositions may fall. They may be waste-waters polluting rivers and coast lines and such as to inhibit treatment processes. They may be liquids or solids, both organic or inorganic, dumped on land sites from which they leach into ground water.

They may be also dumped or flushed directly into the seas, which like the atmosphere and contrary to received wisdom, are not so vast that they are capable of diluting, absorbing and rendering harmless all that reaches them. Whatever the nature of toxic wastes and whatever route they may take, there remains the risk that they may be taken up by life forms, including plants, on land, in fresh water or the seas, and possibly enter our food chains irrespective of their degree of toxicity.

2 EVOLUTION OF UK ENVIRONMENTAL POLICY

Pollution control has a long history in the United Kingdom. In fact, this was the first country in the world to introduce systematic controls over **air pollution** and **waste disposal.** Not altogether surprising - we were after all the first country to undergo an industrial revolution and widespread urbanisation. On 1 April 1996, the UK became the first country, outside the United States, to create a national enforcement agency regulating all potentially harmful wastes from source to final safe disposal or decomposition under a system called Integrated Pollution Control (IPC). It may be helpful to look at how British policy has evolved in response to British concerns and traditions and then from there what impact the European Community has had on these.

3 WATER POLLUTION CONTROL

Water pollution also has a long history of control. 1848 saw the first major piece of legislation in response to public health problems, notably cholera, caused by water pollution. Later, the Rivers Prevention of Pollution Act, 1876 made it a criminal offence to pollute any British river. This Act introduced the concept of "best practical and available means" with respect to **sewage discharges**, and also brought in the first controls on **industrial effluents**. It remained in force until 1951.

Meanwhile the Public Health Acts 1936 to 1937 established a regulatory framework for discharges to public sewers based upon a system of conditional consents for individual discharges. A similar regime was introduced for discharges to rivers by a series of Acts between 1951 and 1974. The 1951 Act also placed a general duty on River Boards and their successors to maintain or restore wholesomeness of rivers. No guidance was given as to how this duty was to be defined or achieved.

Under this regime, normal practice was to relate consent conditions to the volume, nature and use of the receiving waters. This pragmatic approach meant that stricter controls would be imposed, for example, where a river was a source for drinking waters than where it was used only for industrial purposes. No National Water Quality Objectives were produced until the mid-1980's – and then only in response to European Community requirements.

4 THE FORMATION OF THE NATIONAL RIVERS AUTHORITY

During the 1986-1987 Parliamentary Session, the Environment Select Committee conducted an enquiry into the **Pollution of Rivers and Estuaries** and produced a rushed report in May 1987, the day before Parliament was dissolved for the General Election. Subsumed by the ensuing political campaign, it received little media attention, but it did have a significant effect on the policy of the Government and the Water Act, 1989 which followed.

Firstly, it was critical of successive administrations which, between the mid 1970's and early 1980's, had held back the Water Authorities from investment in the renewal of sewage treatment works because of other demands on the Public Sector Borrowing Requirement. In consequence it was found that, in 1986, 22% of all sewage works had failed to comply with the discharge consents for 95% of the time. In one area, failure to carry out timely maintenance at 15 sewage treatment works meant that 73 miles of river length had to be down-graded by at least one Class. This pointed the way to privatisation, since the new Water Companies would be able to borrow directly from the money market for their capital investment needs.

Secondly, the Committee was critical of the fact that the old Water Authorities were both operators and regulators – the poacher/gamekeeper syndrome. They could not prosecute themselves for breaches of their discharge consents, and this in turn inhibited them from prosecuting other polluters; particularly those who could have raised a defence that a stretch of river was more polluted by a prosecuting Water Authority than a factory it had in its sights. Between 1980/81 and 1984/85, out of a total of 64,667 reported pollution incidents in England and Wales there had been only 706 reported prosecutions. This led to the inexorable conclusion and recommendation for the creation of a new unified and independent regulatory body. Two months later, the Government announced its proposal for the National Rivers Authority (NRA) to take over the regulatory functions of Water Authorities upon privatisation.

5 OTHER SOURCES OF POLLUTION

The Select Committee enquiry into Pollution of Rivers and Estuaries had been prompted by their earlier enquiry into **Radioactive Waste**, the Report of which was published at the end of January 1986 and in which concern was expressed at the water flow from the river Drigg through a landfill site alongside Sellafield for the burial of low level waste and into which higher levels could find their way without check. Evidence showed that liquid discharges from Sellafield and Drigg, containing radionuclides, did not necessarily "dilute and disperse" but could be absorbed by sea sediments or become "biochemically" concentrated by organisms and return to land and the human environment.

The Committee revisited Drigg in 1989 and found that carefully engineered concrete bunkers had replaced the open clay trenches and that low level waste, after sampling, was kept there in approved metal containers. Intermediate waste was being encapsulated in cement and high level waste subject to vitrification. Geological explorations have continued for deep disposal of contained intermediate and vitrified, high level waste at sites free from any water flow, but have run into hard opposition from NIMBYs (Not In My Backyard). In 1989, recent and planned expenditure in support of waste management policy amounted to around £1.7 billion supported by a research and development programme of some £20 million a year. Disposal costs at Drigg had risen from £26.80 in 1985-86 to £260 in 1988-89 per cubic metre. However, the discharges of alpha and beta radiation had declined to about 1% of what they were at their peak and monitoring of shellfish confirmed that this had resulted in reduced exposure to the public.

The evidence the Committee received both on River Pollution and Radioactive Waste led them to look at **Toxic Waste** in 1988/9, **Contaminated Land** in 1989/90. The reckless handling of toxic wastes gave the Committee more cause for alarm than in any other environmental enquiry. Although the Control of Pollution Act 1974 had created Waste Disposal Authorities (WDAs), by 1989 only 56 out of 79 of them had submitted their plans to the Department of Environment for the disposal of controlled waste in their areas, apparently without criticism. This laissez-faire attitude resulted in a lack of consistency in standards between one WDA and another. In some, the standards were so low as to pose positive dangers to the public. The findings of this enquiry were confirmed by that which followed into contaminated land where it was found that wastes of all kind were frequently dumped into holes in the ground without monitoring or control nor with regard to leachates into groundwater or gasification of methane into the atmosphere with a global warming effect exceeding that of an equivalent volume of CO_2 by a magnitude of ten. It was little short of a miracle, or the deposits of impervious clay in our subsoil, which had saved the UK from disasters experienced at Love Canal, New York and Lekekerke in Holland. Toxic waste was more insidious than radioactive waste as it could not be as easily detected until it caused actual harm to the environment.

In addition, no effective sanctions existed in respect of on-site commercial or industrial operations contaminating land. Historically, the English Common Law provided some remedy in actions for damages in Tort, where such operations resulted in damage to adjoining land. Such actions reached back to the reign of Henry VII when lime from a tanner's pit polluted the fish stream of the Prior of Southwark. The case law grew piece-meal over the intervening centuries with litigants never sure of which way the precedents would affect them as the Courts evolved differing criteria for nuisance, negligence and strict liability under what is known as the Rule in Rylands & Fletcher. The threads were pulled together in the Cambridge Water Case, decided in the House of Lords in 1994. There a tanner, once again, allowed perchloroethene (tetrachloroethene), a more modern agent than lime, to spill onto his land over a period of years, in the 1960s and '70s, whence it migrated into a borehole of the Cambridge Water Company 1.3 miles away. The solvent was not detected in the drinking water supply until 1983, at levels above those permitted by an EEC Directive issued in 1980 but without any apparent effects upon health. The Water Company won its case in the Court of Appeal and was awarded nearly £1 million in damages, but lost in the House of Lords essentially on a new principle that the tannery could not have foreseen the damage at the time the spillages occurred. The Judges added, echoing a recommendation of the Select Committee, that Parliament should legislate if strict liability for escapes of pollutants was required. However, the NRA subsequently exercised its existing powers and made the Eastern Counties Leather PLC enter into an arrangement for pumping out the chemical still lying at a depth of 50 metres under its land.

The main recommendations of the Select Committee in its Toxic Waste and Contaminated Land Reports are to be found enshrined in the Environmental Protection Act, 1990 and the Environment Act, 1995 encompassing such concepts as a statutory "duty of care" in the production, carriage and final disposal of all waste other than domestic refuse, a system of IPC, and a responsibility upon local authorities to seek out contaminated land in their areas and to require its clean-up where it is likely to cause harm to the environment.

6 THE ENVIRONMENT AGENCY

The last remaining major recommendation to be implemented was the creation of the new Environment Agency which brings together, in one body, all the former regulatory functions of the NRA, Her Majesty's Inspectorate of Pollution (HMIP), and the Waste Authorities – a one-stop agency exercising IPC. Its effectiveness will depend very much on the resources made available to it, but I have no doubt of the determination of Government to see it succeed. One weakness remains; inherited from the former Water Authorities via the NRA, the Agency remains responsible for operational matters such as river navigation and recreation activities, land drainage and coastal defence, water resource management – a continuity of the poacher/gamekeeper syndrome. The Select Committee argued strongly against the Agency taking other than the regulatory functions from the NRA; a position supported by the Royal Commission, the RTPI[A], Institute of directors, the CBI[B], NAWDC[C], CIA[D], but opposed by the CPRE[E], several other conservation groups and the Anglers' Associations. Ministers stated that the Department of the Environment was deeply divided upon the issue. I entered into a public debate with Lord Crickhowell, who wished to keep the NRA empire intact. He took the battle to former Cabinet colleagues and won with particular support from MAFF. However, I do not think the matter ends there and I

[A] Royal Town Planning Institute
[B] Confederation of British Industry
[C] National Association of Waste Disposal Contractors
[D] Chemical Industries Association
[E] Council for the Protection of Rural England

have no doubt that eventually we shall see privatisation of some of the operational functions or the creation of new agencies which will become subject to outside control.

7 THE EUROPEAN DIMENSION

No discussion of these matters is complete without reference to European legislation where there has been much activity in the area of Water Quality and Large Combustion Unit control with draft Directives still in the pipeline for Landfill and Civil Liability for Damage to the Environment. Brussels also has created its own Environmental Agency, but this is a data collecting body rather than the regulatory body which the European Parliament would like to see it become.

Earlier, I mentioned that in the UK no national Water Quality Objectives were produced until the mid-1980's - and then only in response to European Community requirements. It is necessary to add that there has been a long standing debate between the UK and the EC on the theory of how to control routine discharges to water. For dangerous substances, the Environmental Quality Objective (EQO) approach has prevailed in the UK, whereas the Fixed Limit approach has been adopted by all other EC Member States. With the EQO approach, discharges are governed by the concentration of pollutants and their effect on the potential use of the receiving water, after dispersal into the water. With Fixed Limits a uniform limit of pollutant load and concentration is applied to all effluents regardless of dilution. In many cases the Fixed Limit is set near to what it is technically possible to detect. As chemical analysis becomes more effective with the development of more and more sophisticated laboratory apparatus, so the permitted levels of substances tend to be lowered with detectiblity, in compliance with the "precautionary principle", rather than on the basis of proven harm. The "precautionary principle" simply stated is "if you can't prove its harmless, don't allow it" – something our beef farmers have learned to their cost.

The European Commission explained their support for the Fixed Limit approach by their belief that it is easier to enforce and monitor across the community than the EQO approach. This is probably true of continental countries where the Napoleonic Code establishing norms has prevailed. However, the UK has a long tradition of EQO monitoring and our experience and acceptance of the concept of "reasonable conduct having regard to its consequences" has served us well. This is evidenced by the fact that over 90% of our rivers are "good" or "fair" in quality, something not equalled elsewhere in the Community.

8 COMMUNICATION ON EUROPEAN WATER POLICY

Since the above text was prepared, the European Commission has published a Communication on "European Community Water Policy" (Comm. (96) 89) which it seems should lead to the tabling of a new Water Resources Framework Directive before the end of the year. Of immediate interest is the discussion in chapter 7 of the "lengthy debates . . . of the two extreme view points of the 'environmental quality objectives approach' and the 'emission limit values approach' over . . . what is the required level of emission control."

The document goes on to state –
"In practice neither of the two extremes offers an ideal solution. Environmental Quality Objectives alone are often insufficient to tackle serious pollution problems and can be abused as a 'licence to pollute' up to a defined level. Likewise a strict emission limit value approach based on 'Best Available Techniques' (BAT) can in some circumstances lead to unnecessary investment without significant benefits to the environment. More recent Community Legislation takes the emission limit value approach as its point of departure. This is consistent with the precautionary principle and establishes that industry should be responsible for reducing emissions as far as reasonably possible. However there will be occasions when such measures are inadequate to protect the environment and in those circumstances reductions of emissions beyond BAT will be necessary. On the other hand there will be occasions when the environmental benefits of applying the generally applied BAT norms cannot justify the cost. In practical terms the existence of environmental quality objectives allows authorities to judge the effectiveness or otherwise of the emission limit values adopted and whether they need to be tightened. Conversely controls on emissions (usually based on BAT) are the key element of any strategy to ensure compliance with environmental quality objectives. The two approaches are therefore complimentary and not contradictory."

All this denotes a remarkable shift from the established continental view towards the UK tradition - away from absolute limits easy to enforce towards questions of "what harm does it do or how much will it cost to eliminate?"

The communication is of much importance because it extends for the first time to quantitative aspects of water management. It also sets out to define the "Objectives of a Sustainable Water Policy" the challenges presented by point source, diffuse source and accidental pollution, acidification and eutrophication. It dwells on water shortages and the impacts on the water environment of large scale civil engineering works. It seeks to lay down the underlying principles of EC water policy reiterating those of the "precautionary principle", "preventative action" and "the polluter pays". Cost benefits have to be brought into play in considering options.

The outline of the new proposed Directive envisages that eight existing Water Directives remain mostly unaffected but five are to be repealed and replaced. It is intended to establish common definitions for use in all EC water policy and to require integrated water management planning on a river basin basis.

The communication is essentially a consultative document, upon which the comments of all interested parties are invited, and the Framework Directive which will ensue subject to the advice of the Parliament and agreement of the Council of Ministers will undoubtedly make its impact felt with renewed legislation in the Member states for its detailed implementation.

9 THE FUTURE

After having been concerned with the politics of these matters for over ten years, I feel all decisions must be firmly based on sound science. It is important that decisions are not taken because of emotive and populist activity and there must always be a careful cost/benefit analysis of measures to be taken. However, it is not always feasible to delay political decisions until approved laboratory methodology produces an accepted proof. A judgement has to be made on the basis of the balance of probabilities - it is in this situation that the "precautionary principle" has its validity. The problem is that the application of the principle becomes difficult and controversial when scientific consensus is absent. Then an even more difficult judgement has to be made - whether the lack of consensus is the result of genuinely differing interpretations of the facts and not of human rivalry. This is where I believe politicians, who have their own rivalries to contend with, begin to get it wrong.

The Scientific, Economic, Environmental and Political Significance of the Discharge of Toxic Wastes to the Aquatic Environment

D. J. H. Phillips

Fenviron, 87, Chancery Lane, London WC2A 1EU, UK

ABSTRACT

Due to the broad scope of the matters addressed here and the need to focus on a variety of aspects relevant to the significance of toxic waste discharges to aquatic environments, much of the discussion concentrates on a particular topic, which involves the utilisation and discharge of a single class of compounds: the polychlorinated biphenyls (PCBs). It is submitted, however, that this simply focuses the discussion provided, and that the conclusions drawn may also be applied to the utilisation, discharge and toxic impacts in aquatic environments of a wide range of inorganic and organic contaminants.

PCBs are accumulated to considerable extents by aquatic organisms, due to both their lipophilic nature and their resistance to chemical degradation. While the PCB congeners of higher degrees of chlorination generally tend to bioaccumulate to greater extents, the position of chlorine substitution is also important in defining the net uptake of individual PCB congeners by biota. The tendency of PCBs to bioaccumulate and the widespread use of these compounds has given rise to significant contamination of biota worldwide, although this was not recognised until almost 40 years after the introduction of PCBs in industrial applications. Moreover, it has taken a further two decades to adequately characterise the basis for the toxicity of PCBs, which depends principally on the non-*ortho* coplanar congeners. There is a need for improved methods of screening and testing compounds for their environmental impacts, and several such methods are now available, with more in development.

The quantification of the economic effects of PCBs is extremely challenging, due to the uncertainties surrounding potential cause and effect links, and the difficulties in placing monetary values on environmental resources. On the basis of the literature to date, it appears probable that PCBs are exerting detrimental impacts on certain fisheries and on some marine mammal populations, but the costs of this are impossible to estimate presently. Similarly, sublethal effects of PCBs may exist on other species including humans, but the data are not sufficient to estimate this (or its economic value) with any degree of certainty.

With respect to political matters, various problems are identified and discussed herein. These relate mainly to the funding and quality of basic research and training in the aquatic sciences and environmental protection; to efforts concerning international cooperation; and to the use of the Precautionary Principle in deciding on any controls to be introduced for waste discharges. It is concluded that there is little scope for optimism, and that the current political and administrative systems for protecting aquatic environments work poorly. The possibility is raised of commissioning a high-quality national forum (perhaps together with a similar international organisation) to debate such issues and to attempt to provide a direction for future improvements.

1 INTRODUCTION

This manuscript addresses a very wide range of general topics concerned with the impacts of toxic waste discharges on aquatic environments. To attempt to focus the review, the specific discussion provided is mostly restricted to a single class of compounds: the polychlorinated biphenyls (PCBs). The restriction of much of the review to PCBs does not materially affect the conclusions drawn, however, as the history of the use (and misuse) of these compounds provides a case study which may be applied in more general senses to many other aquatic contaminants.

Three main sections are presented below, these dealing with the scientific and environmental significance of PCB utilisation and discharge to aquatic environments; the economic effects of such discharges; and the political aspects of the control of contamination of aquatic ecosystems. The last of these is addressed in a broad sense, with coverage being provided of several aspects of political or quasi-political import. Generic conclusions are provided throughout, these being relevant not only to PCB discharges, but to the utilisation and release of all potentially toxic contaminants. Finally, thoughts are provided as to the future needs to improve the current situation.

2 THE SCIENTIFIC AND ENVIRONMENTAL SIGNIFICANCE OF THE UTILISATION AND DISCHARGE OF PCBs

2.1 Physico-chemical Properties and the Bioaccumulation of Contaminants in Aquatic Environments

The basic chemical structure of PCBs is shown in Figure 1. Some 209 individual PCB congeners may exist in total, these varying in the degree and

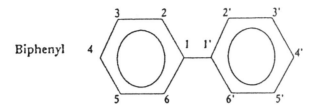

Figure 1 *The chemical structure of the biphenyl molecule.*

Table 1 *The molecular composition of Aroclors 1242 and 1254. ND: Not detected. After Hutzinger et al. (1974).[1]*

	Percentage abundance by weight	
Component	Aroclor 1242	Aroclor 1254
$C_{12}H_{10}$	<0.1	<0.1
$C_{12}H_9Cl$	1.0	<0.1
$C_{12}H_8Cl_2$	16.0	0.5
$C_{12}H_7Cl_3$	49.0	1.0
$C_{12}H_6Cl_4$	25.0	21.0
$C_{12}H_5Cl_5$	8.0	48.0
$C_{12}H_4Cl_6$	1.0	23.0
$C_{12}H_3Cl_7$	<0.1	6.0
$C_{12}H_2Cl_8$	ND	ND

positions of chlorine substitution. However, the environmental abundance of particular PCB congeners is dictated by the fact that these compounds have been used almost exclusively for certain types of industrial applications since their initial introduction over 60 years ago. The commercial mixtures of PCBs which were developed by several manufacturers to support these industrial needs all include a range of individual congeners, and the mixtures available vary in their overall degree of chlorination (Table 1).[1] This reliance on the use of particular mixtures of PCBs defines the congeners of particular abundance in either terrestrial or aquatic environments.

The variations between the commercially-available PCB formulations also affect their physico-chemical properties. While all PCBs are sparingly soluble in water (Table 2),[2-5] the more highly-chlorinated congeners are increasingly hydrophobic, and the precise water solubility of each congener depends not only on its degree of chlorination but also on the positions of chlorine substitution (Table 3).[6] This is of fundamental importance in terms of the bioaccumulation patterns displayed by bulk PCBs (and for those exhibited by other organic contaminants), as the PCB congeners of higher chlorination tend to accumulate to greater extents in biota. Thus, a general correlation has been shown to exist between the water solubility of organic compounds and their bioaccumulation in aquatic systems (Figure 2).[7] Many organisms are thought to accumulate organic contaminants in aquatic environments through a process of lipid-water partitioning,[8] and this effectively explains the correlation

Table 2 *The aqueous solubilities of selected organic contaminants. After Phillips and Rainbow (1993).[2]*

Compound	Type	Aqueous solubility ($\mu g\ l^{-1}$)	Reference
DDT	Organochlorine	17	Biggar et al. (1966)[3]
PCB	Organochlorine	25	Wiese and Griffin (1978)[4]
Dieldrin	Organochlorine	90	Biggar et al. (1966)[3]
Lindane	Organochlorine	2,150	Biggar et al. (1966)[3]
Parathion	Organophosphate	20,000	Martin (1963)[5]
Malathion	Organophosphate	145,000	Martin (1963)[5]
Dichlorvos	Organophosphate	10,000,000	Martin (1963)[5]

Table 3 *Comparative aqueous solubilities of congeners of Aroclor 1242. The concentrations shown are those at day 251 of the experiment. After Lee et al. (1979).[6]*

Peak Number	Number of chlorines	Identity	Solubility ($\mu g\ l^{-1}$)
1	1	2	121.6
3	1	4	15.1
5	2	2,2'	21.2
8	2	2,4'	138.9
9	3	2,6,2'	8.3
10	3	2,5,2'	61.3
16	3	2,4,3'	27.5
19	3	3,4,2'	2.6
24	4	2,4,2',4'	15.9
30	4	2,5,3',4'	7.2
35	4	2,4,3',4'	11.6

noted above.[7] This correlation improves somewhat in relationships between bioaccumulation and octanol-water partition coefficients (K_{OW} values), at least at K_{OW} values below about 10^6 (see Figure 3).[7, 9-11] The octanol-water system has been employed as a surrogate for predicting bioaccumulation in aquatic environments, the partitioning of compounds into the octanol phase being utilised as a model for uptake by biota.

It is important to note that the precise persistence of individual PCB congeners in biological systems depends not only on their overall degree of chlorination, but also on the positions of chlorine substitution. Thus, for example, Tanabe et al. (1982)[12] showed that the biological half-lives of tri-chlorobiphenyls in carp (*Cyprinus carpio*) varied from less than 2 days to more than 90 days, depending on the chlorine substitution pattern (Figure 4). This emphasizes the fact that minor chemical changes in organic compounds can radically affect their potential for environmental damage, and there is clearly a need for systems which can screen chemicals and identify those with particular potential for detrimental effects on terrestrial and aquatic ecosystems.

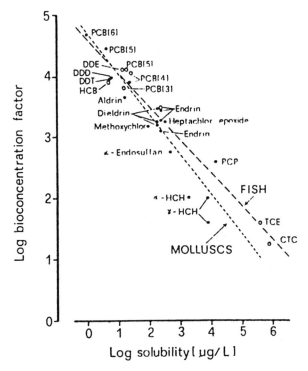

Figure 2 *The correlation between the water solubility of organic contaminants and their bioconcentration in molluscs (filled circles) and fish (open circles). The numbers in parentheses for PCBs indicate their degree of chlorination. TCE is tetrachloroethylene; CTC is carbon tetrachloride. After Ernst (1980).[7]*

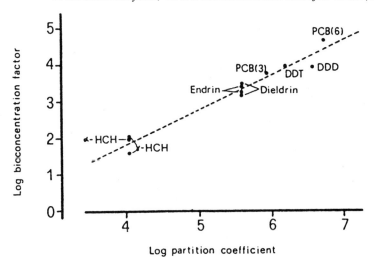

Figure 3 *The correlation between partition coefficients of organic contaminants in the n-octanol/water system and their bioconcentration in bivalve molluscs. Numbers in parentheses for PCBs indicate their degree of chlorination. After Ernst (1980).[7]*

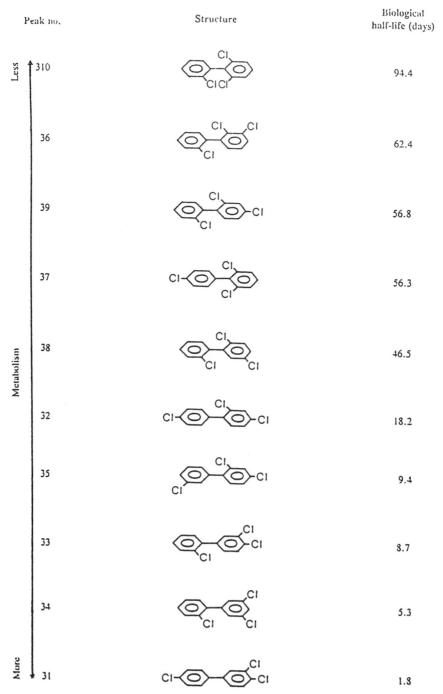

Figure 4 The relationship between the biological half-life of trichlorobiphenyl congeners in carp (*Cyprinus carpio*) and the position of chlorine substitution. After Tanabe et al. (1982).[12]

In this respect, the availability of surrogate models such as the octanol-water system for predicting the probable environmental significance of chemicals is most important. Other systems or concepts have also been developed, including that of fugacity.[13-14] Further work is required, if aquatic environments are to be adequately protected.

2.2 PCB Utilisation and the Recognition of Toxic Impacts

PCBs were first utilised in industry in 1929, and enjoyed a wide range of applications. These include their use as dielectric fluids in electrical equipment such as capacitors and transformers; their utilisation in heat transfer and hydraulic fluids; and applications involving lubricating oils and plasticisers. The electrical applications are of particular note, and environmental reservoirs of PCBs are often associated with ageing electrical equipment.[1]

An example of contamination from the use of PCBs in the electrical industry is shown in Figure 5, and involves the pollution of the Hudson River in New York State, USA.[15] The PCBs were derived from the discharges of the Fort Edward and Hudson Falls facilities operated by the General Electric Company, which manufactured electrical capacitors. The average daily discharges of PCBs (mostly involving Aroclors 1242 and 1016) were computed to be about 14 kg over a 30-year period to the mid-1970s, and this created massive contamination of the entire river catchment downstream of the discharges. Enhanced regulatory controls were introduced in the mid-1970s once the contamintion had been characterised, and the PCB quantities in the discharges had decreased to less than 1 g daily by 1977. This was followed by a reduction in the flux of PCBs downriver over the next decade, which was reflected particularly well in the local biota (Figure 5).[15]

PCB contamination on local or regional scales such as this recurs in many parts of the world, including both temperate and tropical areas. The cumulative tonnage of PCBs in the global environment was estimated by Tanabe (1988)[16] as shown in Table 4. Of about 1.2 million tonnes of PCBs produced to date, only 4% has degraded naturally or been incinerated; some 31% is already present in the open environment, mostly in seawater and sediments; and the remaining 65% either remains in use in electrical and other equipment, or is found in landfills and other sites (most of which are uncontained).[16]

It is clear from these data that PCBs have been very widely used in industry throughout the last five decades. Most importantly, some 37 years passed subsequent to their introduction, before they were initially recognised as potentially important contaminants in aquatic environments.[17-18] Even the initial discovery may not have occurred in the mid-1960s, had not the PCBs been extracted coincidentally by the same clean-up procedures as were employed historically for preparing samples for the study of DDT and other organochlorine contaminants.[18]

This long delay in the recognition of the importance of PCBs in aquatic environments should not be considered unusual. Thus, many other examples

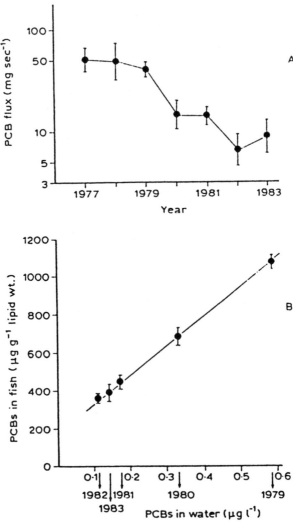

Figure 5 (A) Flux rates of PCBs shown as log values of annual means at Stillwater in the upper Hudson River, between 1977 and 1983. (B) The relationship between mean PCB concentrations in summer samples of yearling pumpkinseed (Lepomis gibbosus) and those in water samples from Stillwater on the upper Hudson River. Vertical lines denote 95% confidence intervals. After Brown et al. (1985).[15]

exist of the slow emergence of an understanding concerning the toxicities of widely-used chemicals. These include the long history of development of comprehensive data on the toxicity of mercury; cadmium; dioxins; and recently, tributyl tin (TBT) formulations employed as anti-fouling agents. In each case, decades passed subsequent to the introduction and widespread use of the chemicals involved, and the eventual realisation of the extent of their environmental impacts.

Table 4 *Estimated loads of PCBs (tonnes) in the global environment, and their percentage contributions to the total. After Tanabe (1988)[5]*

Environment	PCB Load (tonnes)	Percentage of PCB Load	Percentage of world production
Terrestrial and Coastal			
River/Lake Water	3500	0.94	
Seawater	2400	0.64	
Soil	2400	0.64	
Sediment	130000	35.00	
Biota	4300	1.10	
Sub-total	143000	39.00	
Open Ocean			
Air	790	0.21	
Seawater	230000	61.00	
Sediment	110	0.03	
Biota	270	0.07	
Sub-total	231000	61.00	
Total load in the environment	374000	100.00	31
Degraded and incinerated	43000		4
Land-stocked	783000		65
World production	1200000		100

It should be noted that even at present, the direct monitoring of toxicants in aquatic environments involves very few chemicals (e.g. see Table 5).[19] Thus, less than 200 trace contaminants are regularly quantified in most monitoring programmes, of a total of at least 100,000 chemicals in regular use by industry. This led Jernelov (1974)[20] to state the following:

" ... *the total number of compounds to be included in the list of marine contaminants could* ... *be counted in tens of thousands* ... *even out of those which have been looked for and shown to be present, there is only a small number of compounds of which we have more than isolated or scattered data Our ignorance is thus monumental and our ability to evaluate the present situation and future trend regarding total marine contamination is close to non-existent.*"

In an attempt to address this dilemma, the last two decades of research have highlighted the fundamental significance of the physico-chemical properties of chemicals in defining their toxic impacts in aquatic ecosystems. Various test systems have been developed over this period, to improve the screening of chemicals for both their accumulation in and toxicity to biota. It is concluded that there is an urgent need for the further development of both generic and

Table 5 Contaminants for which mussels are analysed in the Californian State Mussel Watch Program. After Stephenson et al. (1986).[19]

Aldrin	Heptachlor
Chlorbenside	Heptachlor epoxide
Cis-chlordane	Hexachlorobenzene
Trans-chlordane	Methyl parathion
Alpha-chlordene	PCB (Aroclor) 1248
Gamma-chlordene	PCB (Aroclor) 1254
Cis-nonachlor	PCB (Aroclor) 1260
Trans-nonachlor	Total PCBs
Oxychlordane	Total phenol
Total chlordane	Pentachlorophenol
Chlorpyrifos	Tetrachlorophenol
Dacthal	Tedion
o,p' DDD	Toxaphene
p,p' DDD	Ronnel
o,p' DDE	Tetradifon
p,p' DDE	
p,p' DDMS	Cadmium
p,p' DDMU	Chromium
o,p' DDT	Copper
p,p' DDT	Mercury
Total DDT	Manganese
Methoxychlor	Lead
Diazinon	Zinc
Dieldrin	Arsenic
Endosulfan 1	Nickel
Endosulfan 2	Selenium
Endosulfan sulfate	Titanium
Total endosulfan	Barium
Endrin	Cobalt
Ethyl parathion	Silver
Alpha-HCH	Tributyl tin
Beta-HCH	Aluminium
Gamma-HCH	
Delta-HCH	

specific systems for screening or testing compounds for their probable bioaccumulation and/or toxicity.

2.3 Recent Data: The Basis for Toxic Impacts

Some two decades passed between the recognition of the existence of PCBs in wildlife in the 1960s and the clarification of the basis for the toxicities of the different congeners. This delay was created by two related factors. The first of these was the difficulty in quantifying PCBs in environmental samples, and specifically in the separation of the individual congeners. The second factor concerned the lack of availability of pure PCB congeners until the mid-1980s, which severely reduced the usefulness of data from toxicological studies.

As the 1980s progressed, these problems were gradually overcome and an understanding emerged of the basis for the toxicity of PCBs in aquatic environments. In the late 1980s, it became clear from a variety of studies that the PCBs of greatest concern were those exhibiting a coplanar structure, and which mimic 2,3,7,8-tetrachloro-dibenzo-p-dioxin. Most notably, these include 3,3',4,4'-tetrachlorobiphenyl; 3,3',4,4',5-pentachlorobiphenyl; and 3,3',4,4',5,5'-hexachloro-biphenyl. It is these three non-*ortho* coplanar PCB congeners which are responsible for much of the toxic impact of PCB mixtures in the environment.[16, 21-24] Later work showed that the non-*ortho* coplanar PCBs are taken up slowly by aquatic biota, but are of exceptional persistence in these species once they have been accumulated.[25] A recent study on PCBs in cod liver oil from the southern Baltic Sea[26] has confirmed that even in areas where bans on the use of PCBs have been in force for many years, the concentrations of non-*ortho* coplanar PCBs have remained quite stable (whilst those of bulk PCBs may have decreased substantially). This is of particular importance in suggesting that regulatory controls on toxic substances are not always successful in reducing the toxicological effects of the compounds involved.[27-28]

3 THE ECONOMIC IMPACTS OF WASTE DISCHARGES

It is widely recognised that environmental economics is only now emerging as a science (or perhaps, art) in its own right, and that the economic effects of many environmental impacts are exceptionally challenging to quantify. Certain aspects of the impacts of PCBs on aquatic environments reflect the difficulties involved, and the most important of these are discussed below.

3.1 The Establishment of Cause and Effect

In the mid-1970s, Helle and co-workers proposed a link between PCB abundance in the Bothnian Bay and the reduced fertility of local populations of the ringed seal (*Pusa hispida*).[29-30] Several confounding and interacting factors were present, however, including elevated levels of DDT and of certain trace metals which correlated to the abundance of PCBs in the system. As a result, no cause and effect link could be adequately established at that time.[29-30] Later research by Reijnders (1980, 1986)[31-32] provided further evidence for effects of PCBs on the reproduction of seals.

Similar findings have been reported which imply a probable link between PCB contamination and the reproduction of finfish. Thus, for example, Spies and co-workers reported suggestive evidence of detrimental effects of PCBs on the reproductive capability of the starry flounder (*Platichthys stellatus*) in San Francisco Bay.[33-34] Later work extended this concept to other fish populations,[35] and although cause and effect links have not been unequivocally proven in these instances, it appears likely that the reproduction of at least some fish species is materially affected in regions which are heavily contaminated by PCBs.

Such difficulties in linking cause and effect have been encountered by most aquatic scientists working in field situations, and unequivocal demonstrations of adverse impacts of contamination have been elusive. Even in the case of the highly specific effects of TBT on oysters and gastropods, significant controversy was generated concerning the basis for the detrimental impacts observed. The consequence of such problems is a long delay in the acceptance of a cause-and-effect link, which inevitably retards the introduction of controls on the utilisation and discharge of toxic substances.

If PCBs are indeed impacting the reproduction (and perhaps other biological aspects) of fish populations and marine mammals in the field, the economic effects of this may be very considerable. However, such potential effects remain largely speculative at present, and improved data are required before any meaningful economic analyses may be completed. There is no doubt, however, that any controls to be introduced on the discharge of toxic substances will have to be largely Government-led, even though the private sector also has a role to play.[36]

The elucidation of cause and effect links for other trace contaminants is also frequently problematic. This is often due to the co-occurrence of pollutants in aquatic systems, with several contaminants exhibiting similar profiles of abundance away from point or non-point sources. In addition, many of the effects of contaminants on biota are non-specific, which tends to frustrate attempts to link cause and effect. A notable exception to the latter involves the impacts of TBT on bivalve molluscs such as oysters and on many species of gastropod mollusc, where highly specific end-points have been noted (the chambering of shells in oysters, and the creation of imposex in gastropods). There is a need for the identification of further such specific effects of contaminants wherever these may occur, as they are of great importance in revealing the causes of biological damage. Researchers in the biochemical fields should bear this in mind, in addition to those studying the biological impacts of pollution at the level of the entire organism. Thus, many so-called biomarkers of contamination appear to be relatively non-specific biochemical responses to stress of various types. These are therefore of little use as indicators of contamination, due to the interfering effects of other stressors in field situations.

3.2 The Potential for Impacts on Human Health

The situation with respect to human health also remains uncertain, at least in relation to the effects of more usual PCB levels encountered in the environment (i.e. in the absence of specific accidental exposures). Much information has been amassed on PCBs in human populations, often through the use of breastmilk which serves as an excellent indicator of total body loads or tissue concentrations.[37] However, the chronic effects of elevated concentrations of

PCBs in humans are uncertain, at least at levels below those generated in accidental high-level exposures. It is intriguing to speculate on the possible oestrogenic impacts of PCBs and on the reported decline in fertility of many species, including the human male in western nations. However, many trace contaminants generate oestrogenic effects, and it is not possible at present to define the source(s) of such impacts. Cause and effect links tend to be even more elusive in studies of humans, due to the obvious difficulties in experimentation.

4 POLITICAL TOPICS: DILEMMAS IN THE SCIENTIFIC AND POLITICAL ARENAS

4.1 Problems and Objectives

Several factors exist which tend to retard progress with respect to the control of waste discharges to aquatic environments. Broadly, these involve the difficulties faced by scientists in providing unequivocal data (and interpretations thereof) to decision-makers; and the need for politicians to balance conflicting demands.

Scientists hold a key position in the generation of both data and debate on the need for controls on waste discharges. It is to be hoped that political decisions concerning matters such as controls on chemical manufacture or utilisation (e.g. such as those introduced during the last two decades in Western European countries for PCBs and TBT) are driven by the initial production of high-quality data, this being subjected to interpretation and peer review, leading to the development of a scientific consensus. This should be followed by an interaction between scientific and political representatives, ultimately generating defensible and appropriate standards on the utilisation of chemicals and on the control of their discharge to aquatic environments. However, it is submitted that this process does not function efficiently for a variety of reasons, as discussed below.

4.2 The Paucity of Funding

Most teaching and research institutions addressing aquatic contamination issues in western nations have experienced a significant decrease in funding levels (in real terms) over the last two decades. The scale of such changes and their ongoing nature threatens the viability of even some of the better-known centres of excellence, with staffing levels threatening to decrease below critical mass in certain instances. The privatisation or quasi-privatisation of many research organisations during this period has detrimentally affected the volume and quality of pure research undertaken.

Neither national nor international governmental bodies appear to have been successful in addressing this problem, and the general level of understanding of this issue frequently leaves much to be desired. It is submitted that little alternative exists to the development of small numbers of high-quality centres of excellence, with adequate levels of funding derived from a reallocation of the existing national and international finance (which is considered to be spread too thinly at present, and to be allocated without a recognition of the over-riding importance of vocational training in the environmental sciences).

4.3 Inadequate Types of Training

The paucity of funding discussed above clearly affects the overall quality of the training available for aquatic scientists. However, a further problem exists which is relevant to the scientific/political interface, and this relates to the type of training provided. It is submitted that debates between most scientists and politicians are doomed from the outset, due to the poorly developed abilities of the two parties to communicate adequately. Few scientists have received training in communication skills, and this severely retards the passage of information from the scientific community to the public and/or politician. By virtue of their training, scientists tend to lapse into technical jargon and caveats concerning probabilities of cause and effect relationships, rather than providing unequivocal statements of "fact" which can be understood and acted upon in the broader public or political arenas.

Political representatives contribute to this impasse, as very few have significant training in the sciences and their abilities to fully understand technical topics and/or the difficulties faced by the scientist in providing unequivocal conclusions are therefore poor. One frequent consequence of this mismatch between the two protagonists is the delegation of both technical and political statements to pressure groups including the so-called Green NGOs, and this can hardly be considered to elevate the quality of the overall debate.

It is considered that the mismatch between scientists and politicians can only be eliminated through improvements in the training of both parties, and this should presumably involve a redirection of at least certain parts of the secondary and tertiary level education system.

4.4 International Co-operation

Much of the international effort related to the control of waste discharges or other impacts on aquatic environments operates under the auspices of the various United Nations (UN) organisations. It is widely recognised that these are under-funded in critical areas of endeavour, and are also massively over-bureaucratised. Scope for optimism with respect to the imminent changes at the top level of the UN Headquarters appears limited, at least at present, and major problems also exist elsewhere in the UN apparatus.

The recruitment policies of the UN organisations are widely considered to be flawed, with political agreements as to preferred geographical and sectoral areas of dominance (e.g. in the Safety Directorate of the International Atomic Energy Agency) and concerns over the mix of country representation in each

UN unit taking precedence over any attempts to optimise the overall quality of the international technical resource. Such overt and covert policies of national Governments severely reduce the drive and usefulness of the international organisations, and this trend appears likely to continue, at least in the absence of a major revolution in the UN and national policies.

Amongst other funding bodies of note, few scientists of quality may be found. The European Union (EU) funds some basic research, but the selection process for funding often appears to be of a somewhat random nature, and the capability of the EU Case Officers to comprehend the data from the studies undertaken is considered to be minimal in many cases.

Controls on the international mobility of scientists are also of concern. While the formation of the European Union is undoubtedly a positive sign in this respect, the difficulties in optimising the mobility of high-quality individuals and in international co-operation involving an exchange of staff (e.g. between Europe and the American continent) have amplified considerably during the last three decades. As a result, many scientific communities are becoming more insular, and their concerns more parochial. While particular conferences provide some outlet for debate of a more international flavour, high-quality events of this type are decreasing in frequency and are exceptionally challenging to fund.

4.5 The Precautionary Principle

Whatever the basic difficulties in identifying cause and effect links between environmental contamination and detrimental effects on resources, a general approach must be developed to decisions on the design and introduction of controls over the sources of such anthropogenic impacts. The existence of the scientific uncertainties noted previously implies that the main issue which arises here relates to the degree to which the Precautionary Principle should be respected in such decisions.

It is sobering to consider that recent scientific publications on the Precautionary Principle have led to controversy and personal animosity, rather than providing an informed high-quality debate.[38-42] Given this performance in the scientific arena in isolation, it appears that an informed consensus from both scientists and politicians on this matter is likely to remain elusive.

4.6 Party Politics or Governmental Co-operation

Scientifically-based debates of importance should experience no input from the polarisation created between opposing political parties. The appalling scenes played out recently in the British Parliament during debates on the Bovine Spongiform Encephalitis problem indicate that this lesson is still to be learned by those who purport to lead the country.

5 TOWARDS THE FUTURE

Recognising the range of problems highlighted here, it appears that there is little scope for optimism. Thus, our accumulated knowledge to date is poor and *"our ignorance is thus monumental"*;[20] the basic funding for research into the effects of waste discharges on the aquatic environment is decreasing; our training is inadequate and probably becoming worse over time, again due principally to funding problems; many scientists and politicians are barely able to communicate, let alone hold an informed debate; international cooperation is poor and will probably become even more difficult in the future; no consensus has emerged on the use of the Precautionary Principle in controlling environmental impacts; and politicians seem intent on constraining debates to party political infighting, rather than addressing the scientific issues of importance.

It is submitted that the general aspect of greatest concern involves the poor quality of much of the debate on many of these general issues of importance to science, and the apparent complete lack of a recognised national and/or international forum to address the problems which must be solved if aquatic (or indeed, terrestrial) environments are to be adequately protected in the future. All scientists, politicians and other authorities must play their part, if improvements are to be made. There appears to be a *prima facie* case for the establishment of a high-quality national forum to debate these issues, preferably coupled to an international arena addressing similar areas of concern.

References

1. O. Hutzinger, S. Safe and V. Zitko, 'The Chemistry of Polychlorinated Biphenyls', CRC Press, Cleveland, Ohio, 1974.
2. D. J. H. Phillips, and P. S. Rainbow, 'Biomonitoring of Trace Aquatic Contaminants', Elsevier Science Publishers Limited, Barking, England, 1993.
3. J. W. Biggar, L. D. Doneen and R. L. Riggs, 'Soil Interaction With Organically Polluted Water', Summary Report, Department of Water Science & Engineering, University of California, Davis, California, 1966.
4. C. S. Wiese, and D. A. Griffin, *Bull. Environ. Contam. Toxicol.*, 1978, **19**, 403.
5. H. Martin, editor, 'Insecticide and Fungicide Handbook for Crop Protection', Blackwell Scientific Publications, Oxford, 1963.
6. M. C. Lee, E. S. K. Chan and R. A. Griffin, *Water Res.*, 1979, **13**, 1249.
7. W. Ernst, *Helgoländ. Wiss. Meeresunters.*, 1980, **33**, 301.
8. J. L. Hamelink, R. C. Waybrant and R. C. Ball, *Trans. Amer. Fish. Soc.*, 1971, **100**, 207.
9. D. W. Connell, and G. J. Miller, 'Chemistry and Ecotoxicology of Pollution', John Wiley & Sons, New York, 1984.
10. G. J. Miller, and D. W. Connell, *Int. J. Environ. Studies*, 1985, **19**, 273.
11. D. W. Connell, *Rev. Environ. Contam. Toxicol.*, 1988, **101**, 117.
12. S. Tanabe, K. Maruyama and R. Tatsukawa, *Agric. Biol. Chem.*, 1982, **46**, 891.
13. D. Mackay, and A. I. Hughes, *Environ. Sci. Technol.*, 1984, **18**, 439.

14 T. Clark, K. Clark, S. Paterson, D. Mackay and R. J. Norstrom, *Environ. Sci. Technol.*, 1988, **22**, 120.
15 M. P. Brown, M. B. Werner, R. J. Sloan and K. W. Simpson, *Environ. Sci. Technol.*, 1985, **19**, 656.
16 S. Tanabe, *Environ. Pollut.*, 1988, **50**, 5.
17 Anon., *New Scientist*, 1966, **32**, 612.
18 S. Jensen, *Ambio*, 1972, **1**, 123.
19 M. Stephenson, D. Smith, G. Ichikawa, J. Goetzl and M. Martin, 'State Mussel Watch Program: Preliminary Data Report, 1985-1986', California Department of Fish & Game, 1986.
20 A. Jernelov, In: *The Sea. Vol. 5: Marine Chemistry*, Ed. E.D. Goldberg, John Wiley & Sons, New York, 1974, p.719.
21 S. Tanabe, N. Kannan, An. Subramanian, S. Watanabe and R. Tatsukawa, *Environ. Pollut.*, 1987, **47**, 147.
22 S. Tanabe, R. Tatsukawa & D. J. H. Phillips, *Environ. Pollut.*, 1987, **47**, 41.
23 S. Tanabe, N. Kannan, T. Wakimoto, R. Tatsukawa, T. Okamoto and Y. Masuda, *Toxicol. Environ. Chem.*, 1989, **24**, 215.
24 S. Safe, *Crit. Rev. Toxicol.*, 1990, **21**, 51.
25 N. Kannan, S. Tanabe, R. Tatsukawa and D. J. H. Phillips, *Environ. Pollut.*, 1989, **56**, 65.
26 J. Falandysz, K. Kannan, S. Tanabe and R. Tatsukawa, *Mar. Pollut. Bull.*, 1994, **28**, 259.
27 D. J. H. Phillips, *Mar. Pollut. Bull.*, 1994, **28**, 192.
28 M. Tester, and D. Ellis, *Mar. Pollut. Bull.*, 1995, **30**, 90.
29 E. Helle, M. Olsson and S. Jensen, *Ambio*, 1976, **5**, 188.
30 E. Helle, M. Olsson and S. Jensen, *Ambio*, 1976, **5**, 261.
31 P. J. H. Reijnders, *Neth. J. Sea Res.*, 1980, **14**, 30.
32 P. J. H Reijnders, *Nature, Lond.*, 1986, **324**, 456.
33 R. B. Spies, and D.W. Rice, *Mar. Biol.*, 1988, **98**, 191.
34 R. B. Spies, J. J. Stegeman, D. W. Rice Jr., B. Woodin, P. Thomas, J. E. Hose, J. N. Cross and M. Prieto, In: *Biomarkers of Environmental Contamination*, Eds. J. F. McCarthy and L. R. Shugart, Lewis Publishers, CRC Press, Boca Raton, Florida, p. 87.
35 J. E. Hose, J.N. Cross, S.G. Smith and D. Diehl, *Environ. Pollut.*, 1989, **57**, 139.
36 B. Garrod, and D. Whitmarsh, *Mar. Pollut. Bull.*, 1995, **30**, 365.
37 A. A. Jensen, *Residue Reviews*, 1983, **89**, 1.
38 J. S. Gray, *Mar. Pollut. Bull.*, 1990, **21**, 174.
39 P. Johnston, and M. Simmonds, *Mar. Pollut. Bull.*, 1990, **21**, 402.
40 A. B. Josefson, *Mar. Pollut. Bull.*, 1990, **21**, 598.
41 J. Lawrence, and D. Taylor, *Mar. Pollut. Bull.*, 1990, **21**, 598.
42 J. S. Gray, *Mar. Pollut. Bull.*, 1990, **21**, 599.

Toxicity Based Criteria for the Regulatory Control of Waste Discharges and for Environmental Monitoring and Assessment in the United Kingdom

J. R. Wharfe

Environment Agency, Guildbourne House, Chatsworth Rd, Worthing, W. Sussex BN11 1LD, UK

ABSTRACT

The effectiveness of using substance specific standards alone for the regulatory control of complex effluents and for environmental quality monitoring and assessment is challenged. Relevant information on the toxicity of many chemicals is either poor or absent, and their fate and behaviour in the aquatic environment is unknown.

The development of scientifically sound environmental quality standards (EQSs) is expensive and, due to the synergistic and additive toxic effects in complex effluents, may fail to provide a guarantee of protection for aquatic life. Additionally, less than 0.1% of listed chemicals have an EQS at present and an alternative approach to control toxic waste discharges is required.

This paper considers the capabilities and limitations of current UK practice and promotes the strategic introduction of toxicity based regulation as an essential part of an integrated approach to control complex waste discharges and to assess environmental quality. Benefits gained from experiences with toxicity based regulation in the United States have greatly assisted research programmes commissioned by the Environment Agency and the Scotland and Northern Ireland Forum for Environmental Research. Much emphasis is placed on the promulgation of methods, and on consistency of approach based on excellence of information. The further development and wider application of Direct Toxicity Assessment is proposed for multi-media waste regulation under Integrated Pollution Control and for consideration of the Best Practicable Environmental Options for waste disposal.

1 INTRODUCTION

It is an offence under Section 85 of the Water Resources Act 1991[1] to cause or knowingly permit poisonous, noxious or polluting matter, or any solid waste matter, to enter controlled waters. Part 1 of the Environmental Protection Act 1990 [2] introduced Integrated Pollution Control which aims to prevent releases of pollutants to air or to make sure where releases do occur that they are kept to a minimum and made harmless.

Current regulatory practice relies heavily on substance specific controls. Environmental Quality Standards (EQSs) derived for some individual substances to protect aquatic life are, in isolation, inadequate to safeguard against damage which may result from the introduction of toxic waste inputs. This is due largely to the small number of determinands for which scientifically sound standards exist, currently less than 0.1% of all listed chemicals, and also because it is very difficult to predict the synergistic and additive toxic effects of combinations of chemicals in complex mixtures.[3,4] Some of the capabilities and limitations of a substance specific approach for regulatory control purposes are shown in Table 1 and are compared with toxicological and biological assessments. The development of whole sample toxicity testing will allow the introduction of Direct Toxicity Assessment (DTA) and adoption of an integrated approach, employing substance specific limits, DTA, and biological survey to provide better control of complex discharges and to assess environmental quality.

In the United States and in Canada, toxicity based criteria have been included in regulatory permits for more than ten years and the approach is gaining wider acceptance in Europe.[5, 6] The development of similar procedures for application in the United Kingdom has drawn on the experience from abroad, particularly from the United States, with emphasis being placed on consistency of approach and excellence of information from toxicity testing.

2 TOXICITY TARGETS FOR REGULATORY CONTROL PURPOSES

The total elimination of contaminants released to the environment and the achievement of pristine quality conditions, cannot be easily achieved, or afforded, by current treatment technologies. Procedures are required to evaluate and manage the risk of environmental damage which might arise from the discharge of toxic waste and to provide meaningful measures of compliance with current legislation. Long term, sustainable environmental benefit gained from reductions in toxic waste, by regulatory control, must be weighed against the necessary investment in treatment processes and realistic timescales of achievement.

Recent studies indicate that many discharges to the environment would breach Section 85 of the Water Resources Act 1991 if a goal of zero toxicity was set at the point of discharge. However, where this happens the environmental damage caused and the cost of amelioration is, in most cases,

Table 1 *Integrated Approach – Capabilities and Limitations*

Control Approach	Capability	Limitation
Substance Specific	Numeric standard can be set	Does not consider all toxics Does not consider all toxic interactions EQS may be analytically difficult to achieve Do not know if EQS is protective
Direct Toxicity Assessment	All toxics are addressed Additive toxicity measured Can predict biological impact Numeric standard can be applied	Incomplete data on causative agent
Biological Assessment	True measure of biological health All stressors measured	Impact has already occurred Cause of impact unknown Cannot set numeric standard Data can be difficult to interpret

unknown. This paper considers a pragmatic approach to progress the better control of complex waste discharges.

The legislation requires a measure of 'poisonous matter', or 'rendered harmless', which relates to the discharge as a whole. Toxicity tests for regulatory control purposes have been, and continue to be, developed. The selection criteria for these tests are described below. The tests should allow a measure of zero toxicity to be established for acute and chronic methods which employ both lethal and sub-lethal end points. These methods must fulfill the selection criteria adopted by the regulatory agency and will effectively set the standard for compliance. The measure of zero toxicity can be selected, depending on the level of toxicity to be controlled, and will vary from acute/lethal to the ultimate goal of chronic sub-lethal no effect. Together with an allowable zone of deterioration, taking account of the available dilution and dispersion in less sensitive receiving waters, the DTA approach provides a set of achievable targets, and timescales for achievement, to be agreed between the regulator and the regulated. The targets are illustrated in Figure 1. Where the toxicity of discharges is unacceptably high, the discharger will be required to undertake a toxicity reduction evaluation programme. The strategic agreement must take account of the need for environmental protection, benefits which should result from reductions in toxic inputs and the means, cost and time of achieving them.

3 THE DEVELOPMENT AND APPLICATION OF TOXICITY TESTS FOR DTA

Toxicity tests suitable for regulatory control purposes need to satisfy a number of criteria. Presented with a bewildering array of test methods, the

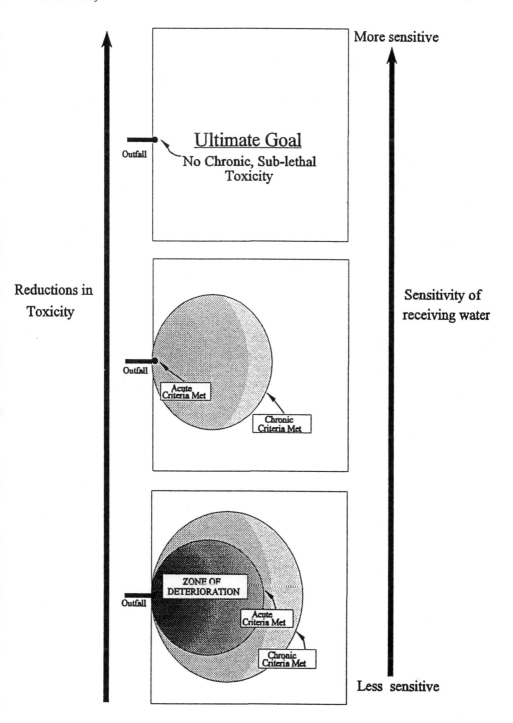

Figure 1 *Zero Toxicity Targets and Points of Application.*

Table 2 Current Toxicity Tests Suitable for Regulatory Purposes

Freshwater	Saline water
Rapid tests:	
Chemical luminescence (e.g. ECLOX^{TM})	Chemical luminescence (e.g. ECLOX^{TM})
Bioluminescence (e.g. MICROTOX^{TM})	Bioluminescence (e.g. MICROTOX^{TM})
Established tests:	
Daphnia sp. (Acute)	Marine copepod (e.g. *Tisbe* sp.) (Acute)
Selenastrum sp. (Acute)	Oyster embryo larvae (Acute/sub-lethal)
Trout (juvenile) (Acute)	*Skeletonema* sp. (Acute)
	Turbot/Plaice (juvenile) (Acute)

Environment Agency has developed an objective system for assigning scores to a number of key criteria which can be weighted for importance according to the specific operational role for which test methods are needed. The seventeen selection criteria used include; sensitivity, reproducibility, repeatability, test method precision, spectrum of response, indigenous test species, ecological relevance and cost.

Application of the selection criteria to many potential candidate toxicity tests, which can be used for regulatory control purposes, identified those shown in Table 2 as being the most suitable for the measurement of acute lethal and sub-lethal toxic effects. Current research initiatives are focused on the development of a battery of rapid tests and the application of chronic sub-lethal tests. In due course it is hoped that some tests might be suitable for on-line capability and others, following further development, for application to land and air environments.

4 EXCELLENCE OF INFORMATION

For regulatory control purposes, it is essential that the information obtained from toxicity tests is of high quality. As with all good analytical procedures, variability in the data, resulting from the test method, must be quantified and controlled. This is best achieved by adoption of methods with clear, unequivocal, Standard Operating Procedures and associated Quality Control. Furthermore, the procedures need to be enforceable through an audit control programme. In cases where these procedures are applied the coefficient of variation on toxicity test methods is equal to and often better than analytical chemistry methods (Personal Communication, M. Rosebrock, North Carolina Division of Environmental Management, North Carolina, US).

Experience in the United States shows that the most successful application of toxicity based permits and reduction in effluent toxicity is achieved in those States with good information control systems. In States with poor control procedures, the data used for regulation have been open to criticism and legal challenge.

Table 3 *Minimum Quality Assurance Criteria*

* Demonstrate competence and experience of staff undertaking toxicity testing
* Adoption of Standard Operating Procedures for all test methods
* Analytical Quality Control records for all test methods
* Facility audit inspection
* Participation in inter-laboratory tests with known samples

North Carolina, one of the more successful State regulators, require dischargers with toxicity based effluent controls to use certified laboratories for testing. The certification scheme run by the State, helps ensure an honest approach to self monitoring with full auditable data being submitted to the regulator. Quality assurance procedures, which include facility inspection and inter-laboratory ring testing, have achieved typical test coefficients of variation of between 20 and 25%.[7]

The introduction of toxicity based criteria for regulatory control purposes in the United Kingdom will follow a similar approach. A performance test exercise, which included regulator based and established test-house laboratories, has been undertaken for four toxicity test methods. This has allowed current levels of performance and coefficients of variation to be assessed. In association with this programme, a scheme to introduce minimum quality assurance procedures has been developed with a view to extending the procedures to full accreditation standard. Some of the criteria under consideration are shown in Table 3. Further details of the scheme and its development can be obtained from the DTA National Centre of the Environment Agency, based in Southern Region, Worthing BN11 1LD.

5 TOXICITY BASED CRITERIA FOR REGULATORY CONTROL

The protocol for toxicity based consenting (Figure 2) has been published elsewhere.[8] A collaborative pilot study, jointly funded by the Environment Agency and the Scottish and Northern Ireland Forum for Environmental Research (SNIFFER), to evaluate the TBC protocol, has identified the levels of toxicity in a range of discharge types. The results will be published elsewhere. Further work is planned to consider the application of chronic sub-lethal toxicity tests and to evaluate protocols for toxicity identification and reduction.

During the development and testing of procedures for toxicity based consenting, the regulatory authorities have taken every opportunity to provide information to the regulated community and to invite comment. A consultation document providing details on the proposals has been prepared and will be widely circulated to seek the views of all interested parties during summer 1996. Key issues for consideration are listed in Table 4. All comments received

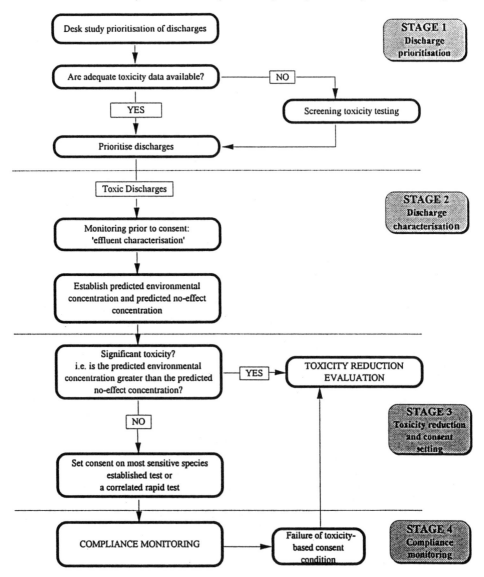

Figure 2 *Proposed Toxicity-Based Consenting Protocol.*

Table 4 *Toxicity Based Consent Consultation Issues*

* Method selection and effluent toxicity characterisation
* Toxicity based consent condition and format of consent
* Sampling programme design and execution
* Data reporting and compliance assessment
* Breach procedures and actions
* Toxicity reduction evaluation and agreements for achievement
* Responsibilities of the Regulator and of the Discharger

will be considered before the operational guidance notes, to allow the introduction of the toxicity based consent conditions, are prepared. Copies of the consultation document will be available (planned release date is July 1996) from the DTA National Centre of Environment Agency based in Southern Region, Chatsworth Road, Worthing BN11 1LD.

6 WIDER APPLICATION OF DTA FOR MULTI-MEDIA QUALITY ASSESSMENT

In addition to their role in regulatory procedures, toxicity tests can be used to assess the general quality of receiving waters and to assess the degree of damage in environmental impact assessments. This will allow toxicity targets in the environment to be achieved and which have ecological relevance. Measures of chronic sub-lethal toxicity, particularly those based on reproductive success or juvenile growth and development, provide a predictive capability for risk assessment by identifying the potential for severe environmental damage well in advance of lethal responses. They can provide more meaningful measures against which investment in treatment processes can be judged.

Current research studies, commissioned by the Environment Agency and SNIFFER, involve the deployment of a range of toxicity tests to identify water and sediment toxicity in both fresh and saline waters, and measures of biological community stress, to further establish their ecological relevance. Some of the toxicity tests selected for evaluation are shown in Table 5. Similar studies, undertaken in the United States have shown good correlation between the predicted in-stream toxicity and the community stress response of macroinvertebrates.[9] The research programme will also consider the use of selected biomarkers to indicate the causative agents of toxicity, although it is recognised at this stage that more work is required to interpret biomarker responses and to establish the credibility of such information based on quality assurance procedures.

The introduction of Integrated Pollution Control, under Part 1 of the Environmental Protection Act 1990, requires prescribed processes involving emissions to more than one medium to be authorised and the Best Practicable

Table 5 Toxicity Tests Selected for Environmental Monitoring and Assessment

Freshwater	Saline water
Rapid tests:	
Chemical luminescence (e.g. ECLOX^{TM})	Chemical luminescence (e.g. ECLOX^{TM})
Bioluminescence (e.g. CHRONITOX^{TM})	Bioluminescence (e.g. CHRONITOX^{TM})
Established tests:	
Daphnia sp. (Chronic)	Oyster embryo larvae (Acute/Sub-lethal)
	Marine copepod (e.g. *Tisbe* sp.) (Chronic)
In situ tests:	
Gammarus pulex feeding rate	*Mytilus edulis* clearance rate
Daphnia sp. (Acute/Chronic)	
Sediment tests:	
Chironomus riparius (Acute/Chronic)	*Corophium volutator* (Acute/Chronic)

Environmental Option (BPEO) for the disposal of potentially harmful industrial materials to be assessed based on Best Available Technology Not Entailing Excessive Cost (BATNEEC). The further development of toxicity tests, suitable for measuring toxicities of land and air based emissions, will greatly enhance this assessment.

At present, there are gaps in the list of available toxicity test methods required to meet the business needs of the Environment Agencies. Some tests, particularly those measuring chronic sub-lethal response and a battery of rapid measures, are still at the development stage, whereas others, including some biomarker tests, need thorough performance evaluation and quality assurance.

The development and implementation of direct biological effect measures, together with the predictive capability of chronic sub-lethal response toxicity tests and the potential of biomarkers to indicate causal agents, will provide more meaningful and easily understood measures for the regulatory control of toxic waste discharges and also for environmental quality monitoring and assessment.

References

1. Water Resources Act 1991, HMSO.
2. Environmental Protection Act 1990, HMSO.
3. National Rivers Authority. 1996. Review of Synergistic Effects in the Environment. R&D Note 461, (Obtainable from The Foundation of Water Research, Marlow. SL7 1FD.)
4. National Rivers Authority. 1996. A Review of Additivity in Toxic Mixtures in the Environment. Draft R&D Note.

5 UNITED STATES ENVIRONMENTAL PROTECTION AGENCY. 1991 Technical Support Document for Water Quality-based Toxic Control. US EPA Office of Water, Washington, DC 20460.
6 DANISH ENVIRONMENTAL PROTECTION AGENCY. 1994 Ecotoxicological Evaluation of Industrial Wastewater. Ministry of the Environment, Denmark.
7 M. M. Rosebrock, N. J. Bedwell, and L. W. Ausley, 1994. N. Carolina Division of Environmental Management, Raleigh, N. C. SETAC 15th Annual Meeting, Denver.
8 J. R. Wharfe and D. Tinsley, *J. CIWEM*. 1995, **9**, 526-531.
9 K. W. Eagleson, D. L. Lenat, L. Ausley, and F. Winborne, *Env. Toxicol. and Chem.*, 1990, **9**, 1019-28.

The Use of Direct Toxicity Assessment to Control Discharges to the Aquatic Evironment in the United Kingdom

D. Tinsley,[1] I. Johnson,[2] R. Boumphrey,[1] D. Forrow[1] and J. R. Wharfe[1]

[1]DTA National Centre, Environment Agency, Guildbourne House, Chatsworth Road, Worthing, West Sussex, BN11 1LD, UK
[2]WRc plc, Henley Road, Medmenham, Marlow, Bucks. SL7 2HDD, UK

ABSTRACT

A protocol for deriving and monitoring compliance with consents containing toxicity criteria has been developed and tested in a pilot study involving UK discharges. The protocol provides guidance on the selection and prioritisation of discharges; the characterisation of effluent toxicity (estimation of the Predicted No Effect Concentration, PNEC); determining the point of application of a zero toxicity target (estimation of the Predicted Environmental Concentration, PEC); carrying out a risk assessment (PNEC v PEC) and thus determining the need for toxicity reduction; setting a toxicity limit; and assessing compliance and enforcing the limit. Both rapid and established acute direct toxicity assessment (DTA) methods were selected for use in the pilot study using selection criteria developed in-house. Twelve discharges were selected from a list of over 60 candidates following the methodology outlined in the protocol. The toxicity of each of the twelve effluents was characterised using the established tests. Testing was also carried out using the rapid tests to look for evidence of a correlation. The results showed the invertebrate tests to be the most sensitive and for ten of the twelve effluents the estimated PEC was equal to or exceeded the PNEC. Toxic effects in the receiving water at the point of application of a zero acute toxicity target would be predicted. In three cases, biological survey data supported the prediction. In the other cases either the data were inconclusive or no data existed. A statistically significant correlation was found between the results from rapid testing and those from testing with the marine invertebrate for two of the discharges. The next steps towards implementation are discussed.

1 INTRODUCTION

The value of using DTA as a part of an "integrated approach" to control the discharge of toxic substances to the aquatic environment in the UK has been discussed in a recent publication[1] and in the previous paper presented at this conference.[2] In the USA, toxicity criteria have been included in regulatory permits for more than a decade[3] and in Europe there has been growing interest in this approach in recent years. In the UK, the need to use toxicity criteria for effluent control was first recognised in the late 60's by the then Ministry of Housing and Local Government. A toxicity test using salmonid fish, derived by the Trent River Authority was recommended for this purpose.[4] However, without a national strategy, without a consistent approach and without investment in developing suitable methods, the subsequent use of DTA in consenting has varied markedly across the UK.

In 1989, privatisation of the water industry took place in the UK and a new non-departmental government body, the National Rivers Authority (NRA), was created to regulate the water environment. The NRA set up a policy group to review the procedures for discharge consenting inherited from the old Water Authorities. As part of this review, the need for toxicity criteria was again considered and a recommendation was made to include these in future consents derived for complex effluents.[5] The job of implementing this recommendation was taken on by a NRA National Ecotoxicology Group.

A protocol was developed based on work carried out at WRc[6] and funding was obtained in a joint venture between the NRA, Her Majesty's Inspectorate of Pollution (HMIP) and the Scotland and Northern Ireland Forum for Environmental Research (SNIFFER) to test this protocol in a pilot study involving selected UK discharges. The objectives of the study were to produce a fully tested protocol for use in controlling potentially toxic discharges. The Water Research Centre (WRc) were commissioned to carry out the work. Support was also obtained to progress other tasks identified as being essential to the successful wider scale introduction of the DTA approach. These included the further research and development of suitable DTA methods; the setting up of a certification programme for toxicity testing laboratories; the provision of DTA awareness seminars and training in DTA consenting procedures for in-house staff, and the creation of a National Centre within the new Environment Agency to support all aspects of the DTA initiative.

In this paper we provide an outline of the pilot study. In particular, we describe the methodology used and discuss the results obtained in terms of the most sensitive tests, the likely environmental significance of the measured effluent toxicity and the evidence obtained for correlations between the results from rapid tests and those from the established algae, invertebrate and fish tests.

2 OUTLINE OF PROTOCOL

The proposed protocol for deriving and monitoring compliance with consents containing toxicity criteria has been outlined in a number of recent public presentations and in recent publications.[1,2] The key stages are summarised in this section. Full details will be provided in a document to be released for public consultation during the summer of 1996.

2.1 Selection and Prioritisation of Discharges

The first stage in the protocol involves the selection and prioritisation of suitable discharges, utilising information on the toxicity of the discharge; a knowledge of its impact on the receiving water; compliance with chemical-specific conditions; the extent of permissive chemical analysis; the sensitivity of the receiving water and the available dilution. It is envisaged that the regulator will carry out the selection and prioritisation process and will expect the discharger to provide the toxicity data, ideally using a battery of rapid toxicity tests.

2.2 Characterisation of Effluent Toxicity

The next stage will determine the Predicted No-Effect Concentration (PNEC) using the most sensitive species identified from testing with established tests representative of three trophic levels (algae, invertebrates and fish). The PNEC is the No Effect Concentration (NOEC) or the EC10 derived from the toxicity testing. The toxicity of each sample tested will also be assessed using the battery of rapid tests with a view to looking for a correlation between the results for at least one of the rapid tests and the test results for the most sensitive species. The use of rapid tests will allow frequent monitoring of effluent toxicity. This should result in a better understanding of effluent variability which in turn should allow the most cost-effective actions by dischargers in reducing toxicity. Rapid tests should also reduce monitoring costs. As with the initial selection and prioritisation of discharges, it is envisaged that the toxicity data required for effluent characterisation will be provided by the discharger.

2.3 Determine the Point of Application of a Zero Toxicity Target

Careful consideration will need to be given to the sensitivity of the receiving water environment and to the available dilution in deciding at which point a zero toxicity target should be applied. If, for example, there is a Site of Special Scientific Interest dependent on water quality near to the end of the discharge pipe and/or minimal dilution in the receiving water, then it may be necessary to achieve a target of zero toxicity at the end of the pipe to avoid unacceptable damage to the receiving water environment. In this case, the Predicted Environmental Concentration (PEC) of the effluent should be considered as 100%. However, if the discharge was to a large estuary with a sizeable dilution, then the benefits of achieving a zero toxicity target at the end of the discharge pipe might not be justified by the costs incurred by the discharger, as water quality in the majority of the estuary would be adequate for the protection of aquatic life. In this case the PEC would be <100%, i.e. an allowance for dilution would be made. It is envisaged that the Environment Agency will seek agreement with dischargers and other interested parties on these points.

2.4 Risk Assessment (PNEC v PEC), Toxicity Reduction and the Setting of a Toxicity Limit

The next stage will compare the PNEC with the PEC. Clearly, if the PEC exceeds the PNEC then the discharger will need to reduce the toxicity of the discharge. This will be achieved through a toxicity reduction programme to be agreed with the regulator. The programme will document the steps that the discharger will take to resolve the toxicity problem, along with timescales. Progress within this programme will be monitored by the Environment Agency. If the PEC does not exceed the PNEC, or once the toxicity of the discharge has been reduced to the required target, a legally-binding toxicity limit will be set for the discharge. This will either specify testing with a rapid test, if a good correlation exists between such a test and the most sensitive established test, or direct testing with the most sensitive established test if a good correlation does not exist.

2.5 Compliance Monitoring and Enforcement

The final stages of the protocol deal with the issue of compliance monitoring which will be carried out by the discharger and audited by the regulator. The regulator will also deal with the necessary action in case of a breach of the toxicity limit in the consent. It is envisaged that compliance monitoring will be carried out using a limit test in which the test result for a single concentration (the PEC) is compared with that of a control. The effluent will pass if there is no effect at the PEC relative to the control. The precision of toxicity tests and the implications for compliance assessment are dealt with in another paper in these proceedings.[7] Such limit tests are used to monitor compliance with toxicity criteria in the USA.[8]

3 SELECTION OF TEST METHODS FOR USE IN THE PILOT STUDY

A number of rapid and established DTA methods were selected from the wide choice cited in the scientific literature. The selection was guided by the application of selection criteria appropriate to the specific operational role of the method as described in the previous paper in these proceedings.[2] Very few of the rapid tests currently available had been fully evaluated. Only the Microtox^{TM} (bioluminescence) and the Eclox^{TM} (chemiluminescence) tests were selected for use as rapid tests in the pilot study.

The selection of established methods was guided by the Organisation for

Table 1 *DTA Methods selected for use in the pilot study*

Freshwater	Marine
Rapid Tests	
– Microtox™ (30 min)	– Microtox™ (30 min)
– Eclox™ (4 min)	– Eclox™ (4 min)
Established Tests	
– Algae (*S. capricornutum*) Lethality (72 h)	– Algae (*P. tricornutum*) Lethality (72 h)
– Invertebrate (*D. magna*) Immobilisation (48 h)	– Invertebrate (*C. gigas*) Larval Development (24 h)
– Fish (*O. mykiss*) Lethality (96 h)	– Fish (*S. maximus*) Lethality (96 h)

Economic Cooperation and Development (OECD) approach of having representative species from each of three trophic levels - algae, invertebrates and fish.[9] As a general principle indigenous species were selected so that the results obtained would be most representative of UK receiving waters. The final selection of tests for use in the pilot study is given in Table 1. All tests used have acute end-points, as established tests with chronic end-points representative of each of the three trophic levels did not satisfy our selection criteria and are therefore the subject of further research and development. Although selected for use in the pilot study, the Environment Agency will reserve the right to add or remove tests from both the rapid and established test batteries for use when the approach is fully implemented.

4 SELECTION OF DISCHARGES FOR USE IN THE PILOT STUDY

Over 60 discharges were put forward for consideration in the pilot study by the collaborating organisations, based on the selection criteria outlined in the protocol. It was necessary to screen a high proportion of these for toxicity as a part of the selection process, since information on their toxicity did not exist. In addition, the biological survey data were less than complete in many cases, making it difficult to decide if some discharges were, in fact, having a deleterious impact on the receiving water.

A number of dischargers were visited by the project team in order to explain the background to the work; the objectives of the pilot study and to discuss the inclusion of their discharge in the study. In each case, a collaborative venture was suggested in which the discharger would provide effluent samples for toxicity testing paid for by the regulator; access to data on the effluent/ industrial process (under a confidentiality agreement as appropriate); sharing of any in-house experience on DTA and the opportunity to meet at intervals to discuss progress. In return, the discharger would receive updates on the DTA initiative and would have the opportunity to influence the development of the approach. In all cases a positive response was obtained from industry and a total of twelve discharges were selected for toxicity characterisation. This final selection consisted mainly of sewage treatment works discharging to freshwater rivers and a range of a industries discharging to estuarine and marine waters.

5 CHARACTERISATION OF THE EFFLUENT TOXICITY OF DISCHARGES IN THE PILOT STUDY

Samples were obtained from each of the twelve discharges at intervals over an eighteen month period and tested for toxicity using both the rapid and established tests. Testing was reviewed after four samples with a view to optimising the use of resources set aside for the study. At this point, it was decided not to carry out further testing with one of the discharges as it was found to have a low and intermittent toxicity. With the other discharges, it was decided not to carry out further testing with algae and fish as the invertebrate test was clearly the most sensitive. In general, a total of eight samples per discharge were tested. In a few cases agreement was reached with dischargers to test further samples either to explore possible correlations between the results from rapid and established tests, or to study the cause of toxicity in a discharge.

Table 2 *Predicted No Effect Concentrations (PNECs) and Predicted Environmental Concentrations (PECs) for the effluents tested in the pilot study*

Discharge	Estimated PNECs from Acute Tests (% effluent)			Estimated PEC (% effluent)
	Algae	Invertebrate	Fish	
A	*	10.0	22.0	33.0
B	*	46.0	*	2.8
C	*	22.0	*	33.0
D	*	4.6	22.0	25.0
E	4.0	0.1	1.0	2.0
F	31.0	0.46	10.0	0.1
G	54.0	0.22	4.6	1.0
H	4.8	0.1	4.6	2.0
I	53.0	0.1	1.0	24.0
G	0.43	0.1	1.0	0.2
K	17.0	0.1	4.6	0.1
L	*	0.1	4.6	2.0

NB: A to D discharge to freshwater rivers
    E to L discharge to estuarine or marine waters
    * = No effects were observed at any of the test concentrations.

Table 3 *Correlations between the rapid and established test results in the pilot study*

Discharge H	Eclox™ IC50 v Invertebrate (*C. gigas*) NOEC r = 0.905 (P<0.01), n = 8
Discharge F	Microtox™ IC50 v Invertebrate (*C. gigas*) NOEC r = 0.815 (P<0.01), n = 8

The lowest PNECs estimated from the test data for each discharge (A to L) are presented in Table 2 along with estimates of the PEC. The results show the invertebrate test to be the most sensitive for both freshwater and marine discharges. The lowest PNECs estimated from the invertebrate tests ranged from 4.6 to 46% for the four freshwater discharges (A to D) and from 0.1 to 0.46% for the eight marine/estuarine discharges (E to L). As discussed in section 2, the significance of these results in terms of environmental quality and the need for toxicity reduction will need to be agreed on a case by case basis. However, a preliminary risk assessment showed the PEC to equal or exceed the PNEC for ten of the twelve discharges and as a result one would expect toxic effects in the receiving water at or beyond the point of application of a zero acute toxicity target. In three cases, biological survey data showed a restricted fauna in support of this prediction. These findings demonstrate the inadequacy of some chemical-specific consents and provides an opportunity for toxicity criteria to demonstrate their value. In the other cases, the available survey data were either inconclusive or no survey data existed.

6 CORRELATIONS BETWEEN RAPID AND ESTABLISHED TEST RESULTS IN THE PILOT STUDY

Evidence of a correlation between the results from a rapid toxicity test and those from the most sensitive established test existed for two of the discharges in the pilot study. Table 3 provides details of the tests, the end-points, the correlation coefficients (r) and an assessment of their statistical significance. Evidence of correlations was also found in a similar study carried out with UK discharges by Johnson et al.[10] Such findings are encouraging and with the addition of further tests to form a battery of rapid tests in future, one might expect to find more of these correlations.

7 NEXT STEPS TOWARDS IMPLEMENTATION

The results of the pilot study and the experience gained provide the Environment Agency with a sound basis for proceeding with the wider introduction of toxicity criteria to help control discharges containing potentially harmful substances. Clear procedural guidance needs to be provided for use by both Environment Agency staff charged with the responsibility of using toxicity criteria to control discharges and by industry who will need to meet the requirements. This guidance is in preparation and will be published for public consultation.

As expected, the pilot study has also helped to identify requirements for further work. This includes the need to develop suitable rapid and chronic sublethal methods, the need to work with dischargers to help them achieve toxicity reduction and the need for further toxicity screening of discharges in order to provide better estimates of the demand for DTA. It is expected that the use of DTA to help control discharges will increase in a phased manner supported by the Environment Agency DTA National Centre. Progress along this path will clearly depend on the success of the approach as judged by improvements in receiving water quality and greater cost-effectiveness.

References

1. J. R. Wharfe and D. Tinsley, J. CIWEM, 1995, 9, 526-531
2. J. R. Wharfe, this volume, p. 26.
3. T. M. Wall and R. W. Hanmer, J. WPCF 59(1), 7-12
4. Trent River Authority Salmonid Toxicity Test HMSO 1969
5. NRA Water Quality Series Report No. 1 1990
6. D. T. E. Hunt, I. Johnson and R. Milne, J. IWEM 1992, 6, 269-277
7. P. Whitehouse, P. A. H. van Dijk, P. J. Delaney, B. D. Roddie, C. J. Redshaw and C. Turner. 1996 Proceedings of Toxic Impact of Waste on the Aquatic Environment. Loughborough University
8. US EPA Technical Support Document for Water Quality-Based Toxic Control. US EPA Office of Water, Washington DC, 1991
9. OECD Guidelines for the Testing of Chemicals - Ecotoxicology. OECD, Paris, 1981
10. I. Johnson, R. Butler, R. Milne and C. J. Redshaw. In "Ecotoxicology Monitoring" (Editor M. Richardson), VCH Publishers, 1993, 309-317

The Precision of Aquatic Toxicity Tests: Its Implications for the Control of Effluents by Direct Toxicity Assessment

P. Whitehouse,[1] P. A. H. van Dijk,[1] P. J. Delaney,[1] B. D. Roddie,[2] C. J. Redshaw[3] and C. Turner[4]

[1]WRc Medmenham, Marlow, Bucks SL7 2HD, UK
[2]ERTL, Leith, Edinburgh EH6 6QV, UK
[3]Scottish Environment Protection Agency, West Region, East Kilbride, Glasgow G75 0LA, UK
[4]Environment Agency, North-East Region, Newcastle NE4 7AR, UK

ABSTRACT

The initiative by the Environment Agencies to control complex and toxic effluent discharges by direct toxicity assessment places great emphasis on the use of aquatic toxicity tests. Like all biological and analytical measurements, determinations of toxicity exhibit variability which can affect both the derivation of the consent condition and also judgements about compliance with the consent. It is important to understand and control this variability if equitable and enforceable regulatory decisions are to be made on the basis of toxicity tests.

This paper reports on the results of a UK ring-test which was conducted to assess the precision of some acute aquatic toxicity test methods which may be used for the assessment and monitoring of complex effluents. It addresses the variability that occurred when repeated tests were carried out on two reference toxicants simultaneously (within-batch variation), within laboratories (repeatability) and also between laboratories (reproducibility).

The implications of these results to the way Toxicity-Based Consents may be set and monitored is discussed. In particular, we consider how test variability may be accounted for when monitoring for compliance with a Toxicity-Based Consent using either conventional concentration-response tests or single concentration 'limit' tests.

1 INTRODUCTION

The initiative by the UK Environment Agencies to control complex and toxic effluent discharges by Direct Toxicity Assessment places great emphasis on the use of aquatic toxicity tests. Like all biological and analytical measurements, determinations of toxicity exhibit variability and this can affect both the derivation of a Toxicity-Based Limit (TBL) to be included in a discharge consent and also judgements about compliance with the consent condition. This paper explains why it is important to control this variability if equitable and enforceable regulatory decisions are to be made on the basis of toxicity tests. We also describe a ring-test conducted in the UK to assess the precision of four 'priority' acute aquatic toxicity test methods and explain how the information generated by the ring-test may be used to set Quality Control standards for acceptable test precision.

2 THE REGULATORY IMPLICATIONS OF VARIABILITY

A TBL is derived on the basis of an investigation into the toxicity of the effluent discharge and the dilution afforded by the receiving water. If a TBL is set, toxicity tests are carried out regularly to monitor compliance with the consent condition (see Tinsley *et al*, this volume). Determinations of toxicity are liable to both random and non-random (systematic) error which will influence the precision of toxicity test data i.e. the degree of agreement between repeated measurements of toxicity made under specified conditions (e.g. within the same laboratory or in different laboratories). This has important implications for the derivation and monitoring of TBLs because it contributes to uncertainty in decision making, as illustrated below.

2.1 Deriving a Toxicity-Based Limit

Variability between laboratories becomes particularly important when a TBL is being derived, because it may affect whether or not consents are set equitably between locations. For example, if the toxicity evaluation of truly identical discharges (with respect to toxicity, dilution in the receiving water etc.) is conducted by different laboratories which typically differ in the sensitivity of the test methods used (perhaps because of systematic differences in the sources of test organisms, dilution water or methods of data analysis), it is quite possible that the predicted no-effect concentrations of the two discharges will be different. As a result, one discharger may be faced with a more stringent consent than the other.

2.2 Monitoring Compliance with a Toxicity-Based Limit

Because of random, and possibly systematic, error most methods of measurement will generate different estimates on different occasions. Toxicity tests are no different. During regular monitoring, it is possible that some effluent samples may be wrongly classified if the errors are substantial: a sample may be regarded as breaching the consent when, in fact, its true toxicity is actually compliant (Type I error or 'false positive') or it might appear to comply with the consent but its true toxicity has been underestimated and it actually breaches the consent (Type II error or 'false negative'). The first should be avoided because of possible adverse commercial implications and the second, because of possible environmental impacts which go undetected.

These uncertainties may be minimised either by making allowances for test variability when making regulatory decisions[1] or by controlling the variability of the test methods used to set and monitor TBLs. Below, we focus on the second approach.

3 THE RING-TEST

3.1.1 Aim. In 1995, a major interlaboratory ring-test of four test methods was carried out in the UK and Republic of Ireland to assess their (a) 'within-test' variability i.e. short-term random error, (b) within-laboratory variability ('repeatability') and (c) between-laboratory variability ('reproducibility'). Using these data, we also propose procedures by which standards for acceptable variability in toxicity test methods may be derived.

3.1.2 Test Methods. Test precision was determined from EC/LC50 values generated by repeat testing of reference toxicants in different laboratories for the following 'priority' methods:

- Microtox™   screening test
- *Daphnia magna* 48h immobilisation test   freshwater invertebrate test
- *Acartia tonsa* 48h lethality test   saltwater invertebrate test
- Pacific oyster (*Crassostrea gigas*) 24h embryo-larval development test   saltwater invertebrate test

Participants were provided with protocols based closely on recognised guidelines (OECD, PARCOM or ICES guidelines) or on the manufacturer's manual in the case of Microtox™.

3.1.3 Test organisms. Because the intention was to measure precision achieved by laboratories under conditions simulating 'normal' practice, the sources of test organisms were not standardised; Microtox™ bacteria and conditioned oysters were obtained from commercial sources whilst *Daphnia* and *Acartia* were usually obtained from in-house cultures.

3.1.4 Test media. Test media were those prescribed in the protocols: Microtox™ tests were performed using commercially available reconstitution medium and *Acartia* and Pacific oyster tests, one of several prescribed sources of clean seawater. *Daphnia* tests were conducted in a synthetic dilution water (Elendt M4 or M7) or suitably treated tap water. EDTA was excluded from the synthetic media when tests with zinc sulphate were conducted.

3.1.5 Toxicants. It was important that the toxicity of the test chemicals should not introduce an additional source of variability. Zinc sulphate, and the moderately water-soluble organic compound, 3,4-dichloroaniline (3,4-DCA), were employed because of their hydrolytic stability. Therefore, exposure concentrations should be readily maintained during experiments and during storage. Furthermore, stock solutions of these reference toxicants were prepared and distributed by WRc, thereby eliminating a possible source of interlaboratory error. Chemical analyses revealed no significant loss of either toxicant when stored between March and December 1995 at 4 °C.

3.1.6 Participating laboratories. Twenty-one laboratories from the UK and Republic of Ireland participated in the ring-test. They represented approximately equal numbers of regulatory laboratories, dischargers (industrial and sewage) and contract research laboratories. All were experienced in the test methods used.

3.1.7 Design of ring-test. The reference toxicants were tested at a range of concentrations from which EC50 or LC50 values could be calculated. Laboratories were asked to perform up to five toxicity tests on different occasions using different batches of test media and organisms but, as far as possible, using the same operator. Provision was made for estimating within-test variability by using a larger number of replicates per treatment than required by the test guidelines, and generating three test datasets using a 'bootstrapping' procedure. The ring-test could be considered as a three-dimensional matrix with simultaneous tests and repeat tests forming one plane and the inter-laboratory comparisons the third dimension.

3.1.8 Analysis of ring-test data. With the exception of data generated using the Microtox™ test, EC/LC50 values were calculated using a moving average angle method. Microtox™ data were analysed using the manufacturer's proprietary software which adopts Michaelis-Menten kinetics to describe the dose-response relationship. In Pacific oyster tests, analysis was based on the normal : abnormal embryo ratio and did not use any assumed or measured inoculum value. Abnormalities in controls were not taken into account but any datasets where control abnormalities exceeded the criterion specified in the protocols (40% abnormality) were excluded. Most tests gave rise to less than 25% control abnormalities.

In practice, the sources of error (within-test, within-laboratory and between-laboratory) which contribute to the measured variability of toxicity tests are not separated but are 'nested' within one another, as illustrated below:

$$\sigma^2_{reproducibility} = \sigma^2_{within\text{-}test} + \sigma^2_{within\text{-}laboratory} + \sigma^2_{between\text{-}laboratories}$$

Residual Maximum Likelihood (REML) techniques were also used to partition the variances into mutually exclusive components, i.e. those on the right hand side of the equation above.

4 RESULTS AND DISCUSSION

4.1 General Description of Test Variability

Table 1 summarises the frequency distributions of EC/LC50 values obtained from all the tests carried out in the ring-test; the results incorporate the combined errors associated with within-test, within-laboratory and between-laboratory sources. Because marked differences in sensitivities between methods were evident, each EC/LC50 was expressed as % deviation from its respective 'consensus' mean (the mean of all results) prior to calculating the distribution parameters in Table 1. It is clear that IC50 values generated by the Microtox™ test exhibited appreciably less variability i.e. individual estimates exhibited less scatter around the consensus mean, than any of the other methods, as reflected by a comparatively narrow interquartile range. When based on responses to 3,4-DCA, tests using *Daphnia* and Pacific oysters resulted in broadly similar levels of precision (as indicated by the interquartile ranges) with the *Acartia* tests exhibiting greater variability. Lower precision was evident in tests when zinc sulphate was employed. In these cases, the 'higher organism' methods yielded similar levels of precision, albeit based on a smaller dataset.

4.2 Sources of Variability

Some understanding of the sources of variability can be obtained from box-plots illustrating the results obtained with each test method by different laboratories (Figure 1). The boxes denote the range of EC/LC50 values from repeated tests (indicating the extent of within-laboratory variability) whilst the 'whiskers' show the range of EC/LC50 values obtained from simultaneous tests (indicating the extent of within-test variability). The values above each box show the number of repeat tests on which the range of EC/LC50 values is based. Only data for 3,4-DCA are shown in Figure 1; zinc sulphate gave rise to comparatively greater ranges when tests were repeated within a laboratory which would contribute to the difference between toxicants noted in Table 1.

It is clear from Figure 1 that within-test variability associated with the Microtox™ test was small but was a larger source of variability for the other

Table 1 *Summary of precision of 'priority' test methods using two reference toxicants*

Test method/Toxicant	No. of tests reported (all laboratories)	Percent deviation of EC/LC50 from 'consensus' mean			
		50%-ile (median)	25%-ile	75%-ile	Interquartile range
Microtox™/3,4-DCA	48	+1.1%	−7.3%	+13.4%	20.7%
Daphnia/3,4-DCA	44	−28.1%	−54.2%	−0.25%	54.0%
Acartia/3,4-DCA	24	−24.8%	−51.0%	+31.5%	82.5%
Pacific oyster/3,4-DCA	31	−16.5%	−40.8%	+17.9%	58.7%
Microtox™/ zinc sulphate	45	+3.4%	−24.4%	+20.4%	44.8%
Daphnia/zinc sulphate	30	+3.4%	−51.1%	+41.6%	92.7%
Acartia/zinc sulphate	15	+8.8%	−58.1%	+45.8%	103.9%
Pacific oyster/zinc sulphate	38	−22.9%	−71.4%	+25.8%	97.2%

methods. This suggests greater control over random error in the Microtox™ test, probably resulting from the highly standardised protocol and high degree of control over reagents and source of test organisms.

The Microtox™ test data also showed small deviations from the consensus mean and comparatively modest within-laboratory variability. By contrast, marked differences between laboratories were evident for some of the other methods, especially in tests using *Daphnia*. The repeatability of *Daphnia* tests was similar in all laboratories but EC50 estimates obtained in Laboratory 30 were consistently higher than those obtained by the remaining laboratories. 'Discordancy' testing[2] showed that the differences were unlikely to have occurred by chance; the *Daphnia* tests in Laboratory 30 may be subject to systematic error, perhaps as a result of the dilution water or genotype of test organisms used. Eliminating this dataset resulted in reduced between-laboratory variability, and precision intermediate between that found with the Microtox™ and saltwater invertebrate tests. Some differences between laboratories were also evident in Pacific oyster and *Acartia* tests but these were not statistically significant, largely because of the small sample size involved and therefore a reduced discriminatory power of the analysis.

REML analysis confirmed the small part played by within-test variability in determining the precision of most of the test methods but it was clearly a greater source of variability in Pacific oyster tests (Figure 2). This was also evident from analyses of tests using zinc sulphate. Increased replication would reduce the impact of this source of variability. For all methods, approximately half the measured variability described in Table 1 was attributable to a combination of errors arising from within-test and within-laboratory sources.

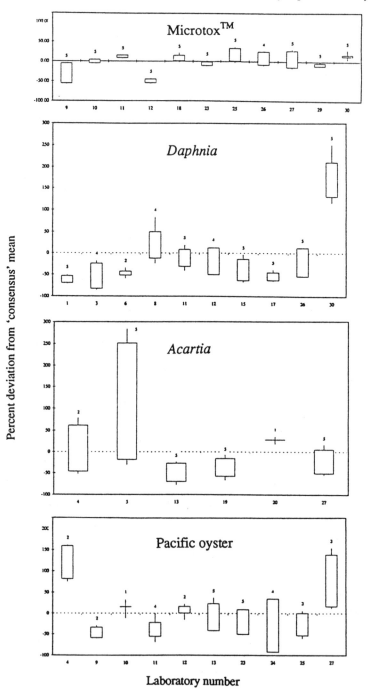

Figure 1 Box-plots showing extent of within-test and within-laboratory variability in different laboratories for tests with 3,4-dichloroaniline (see text for details)

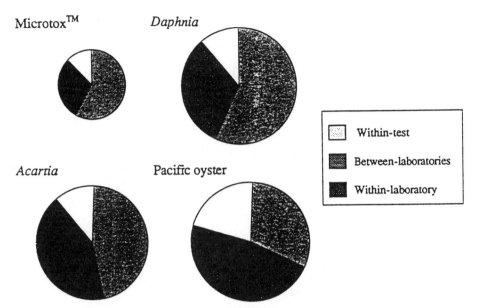

Figure 2 *Standard deviations associated with different sources of error contributing to variability in 3,4-dichloroaniline toxicity (statistical 'outliers' are included). (The area of each pie indicates the standard deviation associated with the reproducibility of the method.)*

4.3 Defining Quality Control standards

The information generated by the ring-test provides an estimate of the 'underlying' variability of the 'priority' test methods. It can also be used to define maximum acceptable standards for precision against which the precision of subsequent tests, such as those carried out to derive or monitor compliance with a consent, can be assessed. We suggest that such judgements should be based on the results obtained from reference toxicant tests carried out in parallel with tests on effluents. Furthermore, because differences in estimates of variability are influenced by the choice of toxicant, the toxicant should be one for which ring-test data have been generated.

4.3.1 Between-laboratory variability. The first step to defining acceptable standards for between-laboratory variability is to remove the influence of statistical outliers. If we assume that log EC50 values obtained by different laboratories assume a normal distribution, appropriate 'discordancy' tests[2] may be applied. Control limits may then be applied to the remaining EC50 estimates, defining the range within which a given proportion (say 95%) of values would be expected to fall. This requires an estimate of the underlying toxicity of the toxicant using that test method (the 'consensus' mean) and the variance for reproducibility, which can be estimated from REML analysis of the ring-test data ($\sigma_{reproducibility}$) is the sum of the variances attributable to within-test, within-laboratory and between-laboratory sources of error):

95% control limit for between-laboratory variability = 'consensus' mean $\pm 1.96 \times \sigma_{\text{reproducibility}}$

4.3.2 Within-laboratory variability. REML analysis of the ring-test data also provided an estimate of the underlying repeatability of each test method ($\sigma^2_{\text{repeatability}}$). This can be compared with variances estimated from the EC/LC50 values derived from a series of repeat tests carried out within a laboratory (S^2) to provide a control scheme using a conventional (χ^2 test, as shown below:

$$\frac{S^2}{\sigma^2_{\text{repeatability}}}(n-1) < \chi^2_{(n-1,\alpha)}$$

If S^2 is significantly greater than $\sigma^2_{\text{repeatability}}$ for that test method, the agreement between repeat tests is less than would be expected based on our estimate of the 'underlying' repeatability of the test, and an investigation of possible sources of error might be prompted to restore within-laboratory variability to an acceptable level.

4.3.3. A possible approach to Quality Control. These approaches may be combined into a Quality Control scheme in which both within- and between-laboratory variability are assessed using EC50 values derived from repeated tests with a reference toxicant (see illustrated example in Figure 3). EC50 values from reference toxicant tests may lie anywhere between the limits for between-laboratory variability but, within those limits, the values should satisfy the criterion for within-laboratory variability. Using hypothetical test data, laboratories A, C and E would meet the criteria for both within- and between-laboratory variability. By contrast, data from laboratory B would be acceptable in terms of the between-laboratory criterion but the poor agreement between EC50 values would fail the criterion for within-laboratory variability. Conversely, data from laboratory D exhibit acceptable repeatability but fail the criterion for between-laboratory variability.

5 CONCLUSIONS

Variability in toxicity estimates using aquatic toxicity tests can introduce uncertainties into regulatory decisions concerned with effluent control by direct toxicity assessment. A ring-test designed to investigate the variability of four 'priority' test methods has revealed clear differences in test precision, with the Microtox™ test being the most precise of those evaluated. With a knowledge of the variability of a toxicity test method, appropriate allowance for test precision may be incorporated into decisions about compliance with a TBL;[1] this study shows that greater allowance would be necessary for tests using *Daphnia*, *Acartia* and Pacific oysters than for Microtox™.

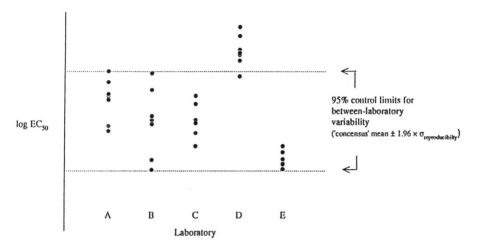

Figure 3 *A proposed approach to judging the acceptability of toxicity test precision (see text for details)*

Within-test variability is small compared to the variability which arises when tests are repeated within a laboratory or compared between laboratories. However, it is a greater source of error with the Pacific oyster test, where its impact may be reduced by increased replication.

Using the ring-test data, we have estimated the 'underlying' variability of these test methods and defined standards against which the precision of tests carried out subsequently, e.g. for regulatory purposes, can be assessed. We propose methods for deriving limits of acceptability for both between- and within-laboratory variability, and show how these may be combined in a Quality Control scheme for aquatic toxicity testing in support of the UK Environment Agency's Toxicity-Based Consents Programme.

Acknowledgements

This work was jointly funded by the Scotland and Northern Ireland Forum for Environmental Research and the Environment Agency. We are also grateful to the laboratories who participated in the ring-test.

References

1 P. Whitehouse, M. Crane, C. J. Redshaw and C. Turner, *Ecotoxicology*, 1996, **5**, 1-14
2 V. Barnett and T. Lewis, 'Outliers in Statistical Data', Wiley, New York, 1978

Amtox™ – A New Concept for Rapid Nitrification Inhibition Testing Applicable to the Laboratory and On-line at Treatment Works

J. Upton[1] and S. R. Pickin[2]

[1]Severn Trent Water limited, Avon House, St Martin's Road, Finham, Coventry CV3 6PR, UK
[2]Engineered Biological Services, 856 Plymouth Road, Slough, Berkshire SL1 4LP, UK

ABSTRACT

A new nitrification inhibition testing system has been developed based on immobilising pure cultures of nitrifiers in a PVA polymer. The system is a continuous flow monitor with blocks (3 mm × 3 mm) of immobilised nitrifiers retained and fluidised in a 1 litre bubble column. The outlet ammonia level is recorded using an ammonia monitor coupled to a data logger.

Inhibition is detected by an abnormal peak in the outlet ammonia level. The threshold inhibition concentration is the point at which the first rise in ammonia occurs. The final level of inhibition is computed by the decrease in ammonia removal before and after addition of the compound under test into the feed.

Trials with a known toxicant, allyl thiourea (ATU), established that the system generates highly reproducible results. At a feed ATU concentration of 1.8 to 1.9 mg l^{-1} the level of inhibition averaged 64.4% with a Standard Deviation (SD) of 1.6%, the response time was 20 to 30 minutes with a threshold of 0.5 to 0.9 mg l^{-1}.

Paired monitors, run in parallel, were directly equivalent. The outlet ammonia averaged 5.1 mg l^{-1} (SD 1.0 mg l^{-1}) and 5.0 mg l^{-1} (SD 0.8 mg l^{-1}) for the two monitors run over five days. When testing a commercially available polyacrylamide marketed for colour removal applications the level of inhibition measured was 29.3% (SD 4.8%) when tested over three days at 5.0 mg active ingredient in the feed. Inhibition persisted, at the same level, three days after the polymer was removed from the feed.

Trials were conducted where the monitor was fed settled sewage to simulate using the system, "on-line", to monitor for inhibition at a treatment works. The system gave stable effluent ammonia levels, averaging 1.2 mg l^{-1} (SD 1.1 mg l^{-1}) with a feed ammonia concentration averaging 37.5 mg l^{-1}, over 44 days. Pre-treatment was required with a column (1.0 litre) filled with reticulated foam blocks to prevent blinding of the blocks. The monitor recorded 100% inhibition by ATU at 10.0 mg l^{-1} with a response time of 1.0 hour.

Immobilisation of nitrifiers enables long term storage (at least six months) at 4°C without significant loss of activity. Replacement cartridges can be stored on site, ensuring minimum downtime and standardisation of the culture.

1 INTRODUCTION

The process of nitrification is well known to be sensitive to a wide range of organic compounds and heavy metals.[1] Waste water treatment works that receive industrial discharges can regularly lose nitrification, due to inhibition, either by an acute shock or by long term, accumulative, toxicity. Consequently, there is a demand for a rapid, laboratory based, toxicity test as well as an on-line monitoring system.

The usual method of testing for inhibition is by flask tests where the culture used varies from a nitrifying, activated sludge sample, from a local treatment works, to a pure culture of nitrifiers. The tests are usually conducted under various conditions of pH, temperature and Dissolved Oxygen (DO) which means that the results are difficult to compare between test runs, and between laboratories. The test procedure is also cumbersome to perform.

Oxygen Uptake Rates (OUR) of pure cultures of nitrifiers, and activated sludge, have also been used as a basis for the measurement of inhibition.[2] However, the disadvantage of measuring the OUR is that it is an indirect measure of the rate of nitrification. It was found that the OUR had a high level of variability between test runs, making an accurate comparison between tests difficult to achieve.

A new method of monitoring for nitrification inhibition is presently under development utilising a culture of nitrifying bacteria immobilised in a Poly Vinyl Alcohol (PVA) matrix. Immobilisation enables the culture to be easily retained in a bubble column and so a continuous flow system can be utilised. As the monitoring system is continuous flow, it can easily be adapted to an on-line monitor, based at a treatment works.

Using immobilised cultures of nitrifiers gives the following key advantages when used as a basis for nitrification inhibition testing:

- Immobilisation enables the same culture, at the same dry solids content, to be used for all monitors, both bench top and on-line. This means that results between tests, and sites, should be comparable.

- The test culture can be stored on site for many months, within a domestic fridge (4°C), without a significant loss of activity, making it easy to retain the test culture on site and replace it following an acute or chronic shock.
- The gel can be cut into small blocks and retained in a bubble column, making a continuous flow monitor easy to run, design and automate.
- A continuous flow system means that the cultures can be kept at the optimum conditions of pH, temperature and DO. This means a stable and high rate of nitrification can be maintained against which differences in rates can be used for inhibition testing. A high rate of nitrification also means that the residence time is low and so shortens the response time.
- A continuous flow system means that both acute and chronic toxicity tests can be conducted, as well as paired tests between treated and untreated columns. A precise estimate of the threshold inhibition level can also be made, based on the onset of inhibition.
- An on-line ammonia monitor can be used, and so automating the test procedure, as a continuous flow of the sample is generated.

Because of these inherent advantages, a development programme was initiated by Severn Trent Water Limited to establish whether a practical monitor can be developed, based on immobilised nitrifiers. The monitoring system is trademarked as "AMTOX"[TM] and the concept of using immobilised cultures for inhibition monitoring has been patented.

This paper reports on the preliminary results from an ongoing programme to develop both a bench top and on-line monitor.

2 MATERIALS AND METHODS

2.1 Method of Immobilisation

A stock culture of nitrifiers was grown on a feed of tap water and ammonium carbonate. The starter culture was from a treatment works that was treating digested sludge liquors and so the mixed liquors had a high proportion of nitrifiers. The culture was grown over a period of five to six months, on a continuous flow basis, and was essentially a pure culture of nitrifiers.

This stock culture was immobilised in PVA. A 500 ml sample of the culture was filtered to give a paste and admixed with 500 ml of the PVA paste (20% w/v). The paste was poured into a shallow tray and frozen at $-20\,°C$ for 24 hours and then allowed to thaw for a further 24 hours. The method of immobilisation was based on the PVA freeze/thaw technique.[3]

The resultant block of immobilised nitrifiers was manually cut into small cubes of a nominal dimension of 5×5 mm. The blocks were washed repeatedly with tap water to remove excess PVA and prevent foaming when aerated in a bubble column.

This method gives a nominal Mixed Liquor Suspended Solids (MLSS) level of the immobilised culture of 15,000 mg l^{-1} of the column. This is estimated from the MLSS of the stock culture and the volume added to the PVA paste. However, it does not take into account evaporative losses during the immobilisation procedure.

2.2 Monitor Details

Although the blocks had negative buoyancy they could be fluidised easily within a bubble column. A relatively large column had to be used to enable sufficient volume of effluent to be generated to pass through an ammonia monitor.

The basic design of the column was a clear UPVC tube of 72.5 cm × 5.5 cm diameter to give a volume of 1.5 litres. The tube was sealed at the base with an end cap with wire mesh glued on the inside of the cap. To this column was added 500 ml of blocks.

The column was filled with tap water and aerated using a small compressor and air stones and heated using an aquarium type water heater with a built in thermostat. The pH within the column was controlled by adjusting the feed to the column to 8.7 (if required) which gave a resultant pH within the column of between 8.0 to 8.3.

Taking into account the volume displaced by the water heater, air and blocks, the resultant water volume was 1.0 litre.

The column was operated with the following conditions:

Feed:	Ammonium carbonate dissolved in tap water to give a feed $N\text{-}NH_4$ of between 14 to 20 mg l^{-1}. Sodium triphosphate was added at 3.0 mg l^{-1} as a phosphate source. The feed was pumped to the column by a Watson Marlow (model number 504 S) peristaltic pump.
Residence time:	30 to 45 minutes
Temperature:	30 °C
pH:	The pH did not require control if the feed pH was maintained at 8.7.

A feed line was taken from the base of the column and fed to an ammonia monitor (ELE Instruments 1801) allowing a continuous reading of ammonia levels within the column. When the ammonia monitor was not used, grab samples from the column were analysed using a spectrophotometer (Hach DR 2000).

2.3 Monitor Stability

To establish the stability of the ammonia removal rate of the system, two columns were run in parallel, with the same feed, as outlined above. The feed and outlet from each column were analysed for $N\text{-}NH_4$, $N\text{-}NO_2$ and $N\text{-}NO_3$ using the spectrophotometer.

The pH, temperature and DO were monitored daily, to ensure that these remained stable between the two columns.

2.4 Response to ATU

In order to measure the response time and accuracy of the system, a series of tests were conducted using allyl thiourea (ATU) as a standard inhibitor to nitrification.

The feed solution was prepared, as before, and pumped to a column containing fully activated blocks of immobilised nitrifiers. The feed ammonia level was measured using a Hach spectrophotometer. For the first four runs, grab samples from the column were taken every 10 minutes and analysed using the spectrophotometer. For two runs, the outlet ammonia level was recorded using the on-line ammonia monitor. A squirrel datalogger was used to record the output from the ammonia monitor, every 15 minutes. The column was fed the untreated feed for 12 hours and then the feed was switched to the same feed, but with ATU added at 2.0 mg l^{-1}. The column was dosed with ATU for a three to five hour period to allow the outlet ammonia level to fully stabilise.

After the test, the feed to the column was changed back to the untreated feed to allow the column to recover fully. The level of inhibition was computed by the reduction in ammonia removal after dosing ATU, expressed as a percentage of the ammonia removal being achieved prior to dosing:

$$\frac{(I - E_u) - (I - E_s) \times 100}{(I - E_u)} = \%\text{Inhibition}$$

Where:

I = N - NH_4 of feed, (mg l^{-1}).
Eu = N - NH_4 of outlet with untreated feed prior to the test run, (mg l^{-1}).
Es = N - NH_4 of outlet with the treated feed, at the end of the run, averaged over the last one hour of the run, (mg l^{-1}).

The same culture could be used for each test, as a full recovery occurred within 10 hours, after changing over to the untreated feed. The threshold inhibition concentration could be computed by calculating the concentration of ATU within the column at the time that the outlet ammonia level first began to rise.

2.5 Response of Polyacrylamide Polymer Used for Colour Removal – Parallel Columns

A polyacrylamide polymer was tested, which was proposed to be used for colour removal, for its chronic toxicity to nitrifiers. The polymer would be dosed directly into the treatment works and there was concern that this polymer may cause a reduction in the rate of nitrification.

Two columns were run, as previously described, and the outlet ammonia levels were allowed to stabilise over a 10 day period. Once stabilised, one column was dosed with the feed containing 5.0 mg active ingredient of the polyacrylamide per litre while the other remained untreated. The treated column was dosed with the polyacrylamide for a three day period, at which point both columns were dosed with untreated feed for a further three days. The outlet ammonia level was recorded, daily, using the spectrophotometer, as was the temperature, pH and DO of each column.

The % inhibition was computed by the decline in ammonia removal of the treated column as a % of the ammonia removal being achieved by the untreated column.

2.6 Simulation of On-Line Monitoring

A preliminary trial in which a column (as previously described) was dosed with settled sewage established that, after a two day period, the blocks blinded with solids. This was caused by the rapid growth of carbonaceous bacteria onto the surface of the blocks. It was, therefore, found necessary to pre-treat the settled sewage using the same design of the column, except that the column was completely filled with reticulated foam blocks of $1 \times 2 \times 2$ cm.

The two columns, run in series, were fed batches of settled sewage obtained from a local sewage treatment works. The total residence time for the two columns was one hour. The inlet and outlet of the columns were analysed for: COD (unfiltered and filtered), $N-NH_4$, $N-NO_2$ and $N-NO_3$. The temperature, pH and DO of the 2 columns were recorded daily to ensure that these were being maintained within the optimum ranges. The trial was run for a 44 day period. The first column began to blind with solids after 14 days and it was found necessary to wash out the excess solids from the foam blocks, every two weeks, to prevent this.

A trial was run to establish the response of this system to ATU. The feed to the two columns was changed to a batch of settled sewage dosed with ATU at 10 mg l^{-1} from the same batch of feed. The level of inhibition was computed as before.

3 RESULTS

3.1 Monitor Stability

The results from the trial where two columns were run, in parallel, with the same feed are outlined in Table 1. It can be seen that the two columns gave a stable rate of ammonia removal with a variation in the outlet $N-NH_4$ of ± 0.3 to 0.4 mg l^{-1}. This was equivalent to a variation of $N-NH_4$ removal of ± 1.5 %.

The total nitrogen level of the inlet was within a standard deviation of that at the outlet, indicating that measuring differences in the outlet $N-NH_4$ accurately reflected changes in the rate of nitrification. No aerobic denitrifica-

Table 1   *Stability of Two Columns Run in Parallel*

		Feed N-NH$_4$ mg l^{-1}	Feed TN mg l^{-1}	Outlet N-NH$_4$ mg l^{-1}	Outlet TN mg l^{-1}	Removal N-NH$_4$ %
Column 1						
	Average	14.4	16.1	1.3	15.7	91.2
	SD	3.1	3.7	0.4	1.1	1.4
Column 2						
	Average	14.1	16.1	1.2	17.4	92.0
	SD	3.1	3.7	0.3	1.4	1.5

Monitor Parameters:

Residence Time: 45 minutes   pH: 8.3
Temperature: 30 °C   DO: 7.8 mg l^{-1}
TN = Total nitrogen
SD = Standard Deviation

tion could be detected with the monitoring system nor significant losses of ammonia by volatilisation.

3.2 Response to ATU

The monitor gave a stable and consistent response to a standard nitrifier toxicant, ATU, at a feed concentration of 2.0 mg l^{-1} (Table 2). The response time was 20 to 30 minutes with a residence time of 35 minutes. Utilising an on-line ammonia monitor considerably eased the running of the test and improved accuracy. A threshold inhibition of 0.5 mg l^{-1} was recorded.

The level of inhibition for ATU at 1.2 mg l^{-1} is quoted as 50%[4] and so the results are within this range, taking into account the higher concentration used for the test.

3.3 Monitor Response to a Polyacrylamide Polymer Used for Colour Removal – Parallel Columns

Two monitors, run in parallel over an extended period of time, showed that the polymer gave a stable inhibition level of 29.3% over the three day exposure period (Table 3) in comparison to the untreated control.

Inhibition continued, even after the polymer was removed from the feed, for a further three day period, with no indication of a recovery. This indicates that the nature of inhibition was chronic and that the increase in the effluent ammonia level, with the treated feed, was not due to the biological degradation of organic nitrogen within the polymer.

This trial illustrates the benefits of using a continuous flow system to detect chronic and accumulative toxicity, which would be difficult to detect with flask based tests.

Table 2 *Monitor Response to ATU at 2.0 mg l^{-1}*

Method of Analysis	Inhibition %	SD %	No. Runs
Colorimetric	70.6	9.1	4
Ammonia Monitor	64.4	1.6	2

Monitor Parameters:
Residence Time: 35 minutes pH: 8.4
Temperature: 30 °C DO: 7 to 8 mg l^{-1}

Table 3 *Response of Monitor to Long-Term Exposure to a Polyacrylamide Polymer Used for Colour Removal*

	Column 1	Column 2	Exposure Days	Inhibition %
	Untreated	Untreated		
N-NH$_4$ Removal %	73.9	74.4	10	–
SD	4.3	3.4		
	Untreated	Polymer (5.0 mg active ingredient per litre)		
N-NH$_4$ Removal %	75.6	55.3	5	29.3
SD	0.6	6.3		4.8
	Untreated	Untreated		
N-NH$_4$ Removal %	76.9	53.7	3	29.5
SD	9.2	1.8		6.1

Table 4 *Simulation of On-line Monitoring with Settled Sewage*

	Feed mg l^{-1}	Effluent mg l^{-1}	Removal %
N-NH$_4$	37.5	1.2	96
Total Oxidised Nitrogen	4.0	23.1	
COD(Total)	416.0	74.0	78
COD(Filtered)	–	53.0	
Suspended Solids	175.0	25.0	

Monitor Parameters:

Total Residence Time: 2.5 hours pH: 8.2
Temperature: 32 °C DO: 7.5 mg l^{-1}

3.4 Simulation of On-line Monitoring with Settled Sewage

Using a pre-treatment column of reticulated foam blocks enabled the monitor to be run with settled sewage, with stable outlet ammonia levels, over the 44 day trial period. The pre-treatment column successfully prevented the blocks of immobilised nitrifiers from blinding with solids. The first column required excess solids to be removed every 10 to 14 days to prevent solids from being carried over into the second column.

If ATU was introduced to the feed at 10 mg l^{-1}, there was 100% inhibition. With a total residence time of 2.5 hours, an increase in the ammonia level in the outlet occurred within 1 hour.

4 CONCLUSIONS

The trials have demonstrated that the AMTOX^{TM} nitrification inhibition monitoring system gives accurate and reproducible results. The system enables both acute and chronic inhibition to be monitored both effectively and routinely.

The AMTOX^{TM} system gives a response time of between 20 to 30 minutes, when used as a bench top monitor, and this would be sufficient for a decision to be made whether a compound, or sample, is acutely toxic. A precise estimate of the level of inhibition would take between 3 to 5 hours, to allow for full equalisation of the response of the monitor.

A key advantage of the AMTOX^{TM} system is that it allows testing of chronic and accumulative inhibition of selected compounds, or samples, to be conducted routinely. The monitor can measure the response to the exposure of a low dose of the test compound over an extended period of time. This means that existing, or proposed, industrial discharges can be tested routinely for chronic inhibition, and a decision taken whether the discharge should be received.

The real potential of the system is to be found in on-line monitoring, and initial trials demonstrate that this is feasible. Future work will concentrate on validating the monitoring system and developing an on-line version.

ACKNOWLEDGEMENT

The development programme is being undertaken by Engineered Biological Services, sponsored by Severn Trent Water Limited. Their continued support is gratefully appreciated.

References

1. C. W. Randall, J. L. Barnard, H. D. Stensel, 'Design and Retrofit of Wastewater Treatment Plants for Biological Nutrient Removal' Technomic Publishing

Company, Inc. 851 New Holland Avenue, Box 3535 Lancaster, PA 17604, USA, 1992.
2 H. D. Stensel, C. S. McDowell and E. D. Ritter, *J. Wat. Poll. Con. Fed.* **48**: 2343.
3 H. Asano, H. Myoga, M. Asano and M. Toyao, *Wat. Sci. Tech.* 1992, Vol. **26**, No 5–6, 1037–1046
4 A. Hooper and K. Terry, J. Bacteriol, 1973, **115**, 480

Performance Trials of a New Automated Respirometer (Merit 20) in Performing Respiration and Nitrification Inhibition Analyses

M. R. Bolton[1], D. Fearnside[1] and R. A. Addington[2]

[1]Yorkshire Water PLC, 2 The Embankment, Sovereign Street, Leeds, UK
[2]A. E. Instruments, Gatherley Road Industrial Estate, Brompton-on-Swale, Richmond, North Yorkshire, UK

ABSTRACT

Respiration and nitrification inhibition tests are becoming more important in assessing the effects of toxic trade wastes on sewage treatment performance. Toxicity testing, of effluents discharged to sewage treatment works, has recently produced evidence that biological treatment processes are severely inhibited in their ability to treat wastes to the criteria specified in their design. Some of the existing protocols are not viable for testing large numbers of samples to acceptable standards of accuracy and precision. In light of this, Yorkshire Water developed a new respirometer in partnership with an engineering firm. The objective was to develop an instrument which could quantify respiration and nitrification inhibition by measuring the respiration rate of the organisms concerned. The instrument should demonstrate an ability to analyse a large number of samples cheaply and to acceptable levels of accuracy and precision. The result of this development work is the "MERIT 20", a twenty channel, manometric, electrolytic respirometer capable of analysing between 9 and 27 samples per day depending on the level of toxicity encountered in the samples. A limited quality control exercise was undertaken on a range of real samples and standard toxic substances. Coefficients of variance ranged between 1.5% and 15% depending on the nature of the sample tested. Further work is under way to establish greater confidence in the protocols. It is envisaged that in the future these tests will be used to monitor and control toxic discharges to STW's.

1 INTRODUCTION

The current respiration inhibition test employed in the Ecotoxicity Section at Yorkshire Water involves the use of a Gilson Respirometer. The inherent labour intensive design of the Gilson, coupled with the fact that it is now out of production, prompted a research and development (R&D) project to design and build a new respirometer to replace existing machinery. In conjunction with Addington Engineers a new 20 channel, computer driven respirometer was designed and built - the MERIT 20 (Manometric Electrolytic Respirometer). This machine underwent a series of trials to assess its precision and reliability in performing the Respiration Inhibition test. Following on from this the current nitrification inhibition test was adapted to run on the MERIT 20, and a series of trials was performed.

2 METHOD

2.1 Respiration Inhibition

The test is an adaptation of the current respiration inhibition method used by Yorkshire Water Services Ltd.[1] Methodology for both tests are based on the Standing Committee of Analysts Blue Book Method[2] and the Organisation for Economic Co-operation and Development (OECD) "Activated Sludge Respiration Inhibition Test".[3] Both tests are designed to quantify inhibition to respiration of activated sludge bacteria caused by the presence of an inhibitive chemical or industrial effluent compared to a control (distilled water) containing no inhibitors.

Measurement of oxygen uptake is carried out by the MERIT 20 automatically. A differential pressure transducer senses the minute pressure changes in the reaction flask, caused by the respiring bacteria. This triggers a steady current to an electrolytic cell which releases oxygen into the reaction flask, restoring equilibrium. Cumulative measurement of the cell 'on-time' indicates the volume of oxygen consumed. The test involves mixing 3.25 ml of sample, 0.5 ml of pre-diluted OECD synthetic sewage, (the concentration of which was 54 ml concentrated OECD synthetic sewage made up to 100 ml with distilled water), and 1.25 ml of activated sludge (at a concentration of 6000 mg l^{-1} suspended solids), from a works receiving predominantly domestic sewage. The mixture is added to a 10 ml reaction vessel and attached to the MERIT 20 for measurement of oxygen uptake rate.

All respiration inhibition calculations were based on measurements taken between the two to three hour contact period. A range of substances were tested, and results expressed as coefficients of variance.

2.2 Nitrification Inhibition

The test is an adaptation of the current nitrification inhibition method used by

Yorkshire Water Services Ltd.[4] Methodology for both tests are based on the Standing Committee of Analysts Blue Book Method[5] and the Nitrotox Toxicity Test.[6] The test uses a commercially available culture of nitrifying bacteria and is designed for use on the MERIT 20. The method described assesses the inhibitory effect of a test substance on a culture of nitrifying bacteria by measuring the respiration rate of the bacteria under set conditions, compared to that of a control.

The test involves mixing 3.35 ml of sample, 0.2 ml of buffer (250 g l^{-1} $KHCO_3$) 1 ml of ammonium chloride solution (1000 mg l^{-1}), 0.4 ml of Biolyte solution (diluted 50% with distilled water) and 0.05 ml of distilled water in a 10 ml reaction vessel. To account for contamination of the nitrifying bacteria by non-nitrifiers, a duplicate sample is simultaneously analysed the 0.05 ml of distilled water being replaced with 0.05 ml of 2000 mg l^{-1} allyl thiourea (ATU) to give a, final concentration of 20 mg l^{-1} ATU. This is a known inhibitor of nitrification. The respiration rate due to the presence of contaminating bacteria can thus be compensated.

The reaction vessel is attached to the MERIT 20 for measurement of oxygen uptake rate. All respiration inhibition calculations are based on measurements taken between the two to three hour contact period. A range of substances were tested, and results expressed as coefficients of variance.

3 RESULTS

3.1 Expression of Results

Inhibition is expressed in two ways, percentage inhibition (% inhibition) and EC50. All results are calculated on respiration rates measured between the two to three hour contact time.

3.1.1 Percentage Inhibition (% Inhibition). The amount of inhibition produced by a test substance compared to a non-inhibitory control, calculated as follows:

$$\% \text{ Inhibition} = \frac{[R_B - R_S]}{R_B} \times 100$$

Where : R_B is the respiration rate of the control.
          R_S is the respiration rate of the test substance.

3.1.2 EC50. The concentration of a substance which produces a 50% reduction in respiration rate. It is calculated by plotting a minimum of three % inhibitions at varying concentrations. Where the slope crosses the 50% inhibition level defines the concentration at which 50% "effect" has occurred.

Performance of the MERIT 20 is measured in terms of the coefficient of

Table 1 *Respiration Inhibition – Synthetic Sewage Blank Run Results*

	RUN 1	RUN 2	RUN 3
Mean Activity	248	250	354
SD	9.3	10.1	12.5
CV	3.8	4.0	3.5

Table 2 *Respiration Inhibition - Domestic Sewage Blank Run Results*

	RUN 1	RUN 2
Mean Activity	358	356
SD	15.7	12.9
CV	4.4	3.6

variance (CV) of the respiration rate for the blank runs and either % inhibition or EC50 results for the test runs. The CV relates the size of variation to the average of the figures it was derived from, enabling the comparison of variability for different tests and methods. Mean and standard deviation (SD) are also given in the results tables. All sets of data are regarded as samples of a bigger population.

3.2 Respiration Inhibition Results

3.2.1 Synthetic Sewage Blank Runs. The blank run results are shown in Table 1. A run consisted of 20 identical samples, using distilled water as the test material, along with synthetic sewage and activated sludge. Results are expressed in μh^{-1} oxygen consumption, readings being taken between the second and third hour.

Overall the MERIT 20 showed a channel failure rate of 3.3% (a failure deemed to be a result differing by more than three times the standard deviation from the mean).

3.2.2 Domestic Sewage Runs. The domestic sewage run results are shown in Table 2. A run consisted of 20 identical samples, using crude sewage from a works receiving predominantly domestic sewage as the test material. Results are expressed in $\mu l\ h^{-1}$ oxygen consumption, readings being taken between the second and third hour.

Overall the MERIT 20 showed a channel failure rate of 10.0% (a failure deemed to be a result differing by more than three times the standard deviation from the mean). Respiration rates were higher than the water blanks, the reason for this being due to the crude sewage containing large amounts of readily available BOD, giving a higher initial respiration rate. The OECD

Table 3 *Respiration Inhibition – Municipal Sewage Run Results*

	RUN 1	RUN 2	RUN 3	RUN 4
	35.0	35.7	34.7	31.6
	33.1	36.0	32.9	31.4
	36.7	36.0	32.1	33.3
	31.7	39.1	32.1	37.6
	36.7	33.8	37.3	34.0
	35.7	36.1	33.4	31.3
	35.4	37.2	31.1	35.1
	37.1	36.7	33.6	35.6
Mean	35.2	36.3	33.4	33.7
SD	1.9	1.5	1.9	2.3
CV	5.4	4.1	5.8	6.8

Table 4 *Respiration Inhibition – Standard Toxicant Run Results*

	RUN 1	RUN 2	RUN 3	RUN 4	RUN 5	RUN 6	RUN 7
EC50 1	8.5	9.7	11.4	11.4	8.4	9.6	9.6
EC50 2	8.8	9.9	11.1	11.5	7.9	9.7	8.7
EC50 3	8.4	10.1	10.1	10.1	8.4	9.4	9.8
Mean	8.57	9.90	10.87	11.00	8.23	9.57	9.37
SD	0.21	0.20	0.68	0.78	0.29	0.15	0.59
CV	2.43	2.02	6.26	7.10	3.51	1.60	6.26

NOTE – A second batch of DCP was used for runs 3-7. Between run CV = 11.15 (using the EC50 values).

synthetic sewage is designed for a five day BOD test and thus has less readily available BOD.

3.2.3 Municipal Sewage Runs. This consisted of primary tank effluent from a works receiving a significant amount of trade effluent. The same sample was used throughout the study. The results are shown in Table 3. Results are expressed as % inhibitions and are calculated using a mean of two results.

3.2.4 Standard Toxicant Runs. The 3,5-Dichlorophenol (DCP) run results are shown in Table 4. A run consisted of two water blanks and 18 DCP samples with final concentrations of 5 mg l^{-1}, 10 mg l^{-1} and 20 mg l^{-1}. Analysis was performed in duplicate giving a total of three EC50 values per run. Results are expressed in mg l^{-1} DCP.

3.2.5 Trade Effluent Runs. The trader chosen for the study had an effluent originating from the manufacture of organic compounds. Table 5 shows these results as EC50 values, calculated using three dilutions of the effluent and expressed as final dilutions.

Table 5 *Respiration Inhibition - Trade Effluent Run Results*

	RUN 1	RUN 2	RUN 3
EC50 1	12.0	12.2	16.0
EC50 2	12.7	12.8	16.6
EC50 3	12.5	12.8	16.6
Mean	12.4	12.6	16.4
SD	0.36	0.35	0.35
CV	2.91	2.75	2.11

Between run CV = 14.31 (using EC50 values)

Table 6 *Nitrification Inhibition – Synthetic Distilled Water Blank Run Results*

Run Number	CV Blank	CV ATU Spike	CV Corrected Blank
1	9.4	9.2	11.2
2	13.2	12.7	18.6
3	9.7	17.1	10.7
4	12.6	2.4	14.6

Analysis was based on one sample over a period of three days, thus degradation of the sample would affect its inhibitive qualities.

3.3 Nitrification Inhibition

3.3.1 Distilled Water Blank Runs. The results for the distilled water blank runs are shown in Table 6. Four runs of 20 blanks were performed, each run consisting of 10 water blanks and 10 water blanks spiked with ATU. Results are given for the blanks, the ATU spiked blanks, and the blanks which have been corrected for the contaminating respiration, all expressed as CV.

Variability increases in going from the blank results to the corrected blank results. This is due to the additional variability of the ATU spike.

3.3.2 Domestic Sewage Blank Runs. The results for the domestic sewage blank runs are shown in Table 7. Four runs of 20 blanks were performed, each run consisting of 10 sewage blanks and 10 sewage blanks spiked with ATU. Results are given for the blanks, the ATU spiked blanks, and the blanks which have been corrected for the contaminating respiration, all expressed as CV.

The sewage blanks gave either no or slight inhibition when compared to water blanks. The ATU spiked sewage samples gave consistently higher activity values than the ATU spiked water samples, this being attributed to the non-nitrifying bacteria present in the sewage itself.

Table 7 Nitrification Inhibition – Domestic Sewage Run Results

Run Number	CV Blank	CV ATU Spike	CV Corrected Blank
1	3.2	13.5	6.4
2	6	17.5	12
3	12.3	9	14.9
4	9	8.7	10.7

Table 8 Nitrification Inhibition – Municipal Sewage Run Results (expressed as % inhibition)

	RUN 1	RUN 2	RUN 3	RUN 4	RUN 5	RUN 6
	52	16.8	12.8	28.3	18.6	23
	75	0	3	12.1	20.2	24
	69	29	23.6	11.1	7.3	11.2
	68	20	12.4	1.3	18.3	0
	66	33	22.2	22.4	9.4	12
Mean	66	19.76	14.8	15	14.8	14
SD	8.5	12.8	8.4	10.5	5.9	9.85
CV	12.9	64.8	56.7	70	40.3	70.2

3.3.3 Municipal Sewage Runs. This consisted of primary tank effluent from a works receiving a significant amount of trade effluent. The same sample was used throughout the study, a total of 5 inhibition values being obtained per run. Results are shown in Table 8, expressed as % inhibitions. Initially, the neat sample proved too toxic (Run 1), and had to be diluted (2 to give % inhibitions below 50%.

3.3.4 Standard Toxicant Runs. The 3,5-DCP run results are summarised in Table 9. The MERIT 20 is capable of giving two 4-point EC50 values per run. A total of 15 EC50 values were obtained over a period of two weeks.

3.3.5 Trade Effluent Runs. The trade effluent results are shown in Table 10. The trader used for this study had an effluent originating from the manufacture of organic compounds. The MERIT 20 is capable of giving two 4-point EC50 values per run. All analysis was performed on the same sample over a period of four days, results expressed as final dilutions.

Table 9 Nitrification Inhibition – Standard Toxicant Run Results

Mean EC50	4.5
SD	0.6
CV	13.2

Table 10 *Nitrification Inhibition – Trade Effluent Run Results*

Run Number	EC50 Trader (Final dilution)
Run 1	24
	20
Run 2	26
	32
Run 3	28
	29
Run 4	27
	24
Run 5	25
	30
Mean EC50	26.5
SD	3.47
CV	13.1

4 DISCUSSION

4.1 Respiration Inhibition

Overall, the MERIT 20 performed well in all of the test runs, achieving a standard of precision comparable to the Gilson respirometer. Although the MERIT 20 did not show appreciably greater precision than the Gilson for the respiration inhibition test, the actual setting up and running of the MERIT 20 proved to be far less labour intensive and time consuming than the Gilson.

Preparation of the MERIT 20 proved to take approximately a quarter of the time required for the Gilson respirometer. This was achieved by time and labour savings in the three key areas of machine preparation, glassware preparation and machine loading/unloading. Preparing the MERIT 20 involved programming the on-board computer via the in-built keypad. Preparation of the Gilson involved resetting each valve individually, removing old grease and applying new grease and topping up the waterbath. Glassware preparation for the MERIT 20 reaction flasks required a two hour soak in 10% Decon then a normal wash/rinse cycle (along with the glass stirrers) in a glasswasher. The Gilson flasks required firstly the removal of the paper wick and the rinsing out of the potassium hydroxide well in the middle of the flask, then an overnight soak in 50% Decon to remove the sealant grease from the neck of the flask. Furthermore, the Gilson glassware was not suited to an automatic wash cycle thus each flask had to be manually rinsed in three times in tapwater then three times in distilled water. A simple screw fit attached the reaction flasks to the MERIT 20, whereas the Gilson flasks had to be attached via two elastic bands, ensuring that the grease had made an air-tight seal.

Pipetting sample and reagents into the MERIT 20 reaction flasks proved to take approximately half the time required for the Gilson and required less manual dexterity. Concentrated potassium hydroxide had to be pipetted into the centre well of each Gilson reaction flask and great care was required in ensuring none was deposited either on the neck of the flask or in the actual sample well. The MERIT 20 reaction flask did not require this procedure and the wide mouth of the flask enabled rapid pipetting of sample and reagents.

Starting and running a test on the MERIT 20 involved choosing the start procedure from the main menu. Once started the MERIT 20 required no attention, the on-board computer monitoring each channel for the duration of the test. This enabled overnight runs to be performed. In comparison, the Gilson respirometer needed hourly attention to manually wind down each valve and record the oxygen consumed. Care had to be taken that the manometer fluid level did not drop too far between readings as an air bubble would form, thus constant attention was required for sample runs which showed a high oxygen demand.

4.2 Nitrification Inhibition

The draft method showed promise with the standard toxicant and trade effluent trials, achieving a good standard of reliability. Municipal sewage trials however gave unreliable results, and the water and sewage blank trials showed appreciable variance as well. The precision and reliability of the test was compromised by the contaminated culture of nitrifying bacteria. Although the contamination was accounted for by using ATU spiked blanks, the effect was to reduce machine throughput by 50% and increase errors. The fact that nitrification was being measured indirectly by oxygen consumption meant that there was a considerable risk of misleading results.

The MERIT 20 again proved to be less labour intensive than current apparatus, however as throughput had been effectively halved due to the contaminated culture, there was little difference in throughput between the two methods.

5 CONCLUSIONS

5.1 Respiration Inhibition

The MERIT 20 shows comparable performance to the Gilson respirometer regarding reliability of results.

The MERIT 20 surpasses the Gilson in terms of throughput, labour costs and data capture.

The MERIT 20 is to replace the Gilson respirometer and become the apparatus for the respiration inhibition test.

The Respiration Inhibition test using the MERIT 20 will be submitted to the Standing Committee of Analysts for inclusion as a Blue Book Method.

5.2 Nitrification Inhibition

The draft method performed on the MERIT 20 did not give results to the desired level of reliability.

The contamination of the pure nitrifying bacterial culture reduces throughput by 50% compared to the Respiration Inhibition test.

Although showing some promise, the method in its present form did not pass the trials.

Further work needs to be done to identify fully the source(s) of error in the draft method before it can replace the current method used by Yorkshire Water.

References

1. D. Fearnside, Respiration Inhibition Standard Operating Procedure, 'Assessment of the Toxicity of Chemical Substances and Industrial Effluents to Mixed Cultures of Organisms Found in Biological Treatment Processes' Yorkshire Water Services Ltd, 1991.
2. Methods for assessing the Treatability of Chemicals and Industrial Waste Waters and their Toxicity to Sewage Treatment Processes, *HMSO* London 1982.
3. Organisation for Economic Cooperation and Development (O.E.C.D.) Guideline For Testing of Chemicals, No. 209, "Activated Sludge, Respiration Inhibition Test", (1984).
4. D. Fearnside, Nitrification Inhibition Standard Operating Procedure, 'Assessment of the Toxicity of Chemical Substances and Industrial Effluents to the Nitrifying Bacteria, *Nitrosomonas* and *Nitrobacter* Found in Biological Treatment Processes.' Yorkshire Water Services Ltd, 1991.
5. The Assessment of the Actual and Potential Nitrifying Ability of Activated Sludge (Tentative Methods) *HMSO* London, 1980.
6. Nitrotox[TM] Toxicity Test, Protocol No. 31, International Biochemicals, Sandyford Industrial Estate, Foxrock, Dublin 18, *R & D Protocols*, 1992 **(Issue 2)**, 5pp.

The Potential for Biosensors to Assess the Toxicity of Industrial Effluents

J. G. Rogerson[2], A. Atkinson[1], M. R. Evans[2] and D. M. Rawson[1]

[1]Research Centre, University of Luton, 24 Crawley Green Road, Luton, Bedfordshire LU1 3LF, UK
[2]Brixham Environmental Laboratory, ZENECA Limited, Freshwater Quarry, Brixham, Devon TQ5 8BA, UK

ABSTRACT

At Brixham Environmental Laboratory, in collaboration with the University of Luton, electrochemical transducers have been used in combination with mediators to interrogate living bacterial cells in a biosensor configuration. This work considers the use of biosensors for toxicity assessment, with particular application to industrial site and process effluents. There are many advantages of this approach over the conventional bioassay methods, in particular, the short test period, ease of use, and the use of appropriate test species. Preliminary results show that the toxicity data generated using the biosensors correlated significantly with those data obtained from appropriate conventional tests such as Microtox™.

1 INTRODUCTION

The need for tests that could provide early warning of environmental threat, has lead to microbes becoming increasingly accepted as test species for use in toxicity testing.[1-9] Although luminescence has been exploited for use in rapid toxicity testing,[10-18] a limited number of bacterial species are employed. Biosensors incorporating bacterial cells enable a wide spectrum of species, and consortia, to be used for rapid ecotoxicity assessment.[19] Incorporation of bacterial cells into biosensor configurations based on electrochemical transducers allows real-time monitoring of metabolic status and the opportunity for rapid detection of cellular perturbation due to toxic challenge. The redox mediator, potassium hexacyanoferrate (III), is suitable for monitoring bacterial[20, 21] and cyanobacterial cells,[22, 23] and biosensor signal reflects the metabolic status of the immobilised bacterial population.

Unlike enzyme and antibody based biosensors, the integrating and recognition capabilities of living cells enable their use in broad spectrum devices for determining complex variables. Biosensors incorporating living cells have been shown to have an application for monitoring environmental pollutants.[19, 20, 22-28] This present work considers the use of biosensor techniques for toxicity assessment with particular application to industrial site and process effluents obtained from various ZENECA businesses. Data obtained will be compared with those obtained from existing test methods.

End user opinion was sought in developing the biosensors, and the features considered most important for such test systems were: environmental relevance of the biological component; sensitivity; ease of use; rapidity; and cost effectiveness. This has led to the development of a generic mediated amperometric approach based on low cost screen-printed electrodes capable of accommodating, and monitoring the metabolic status of a wide range of bacterial biocatalysts.

2 MATERIALS AND METHODS

2.1 Microorganisms

Escherichia coli (NCIMB 8277) and *Pseudomonas putida* (NCIMB 9494) were obtained as freeze dried cultures from the National Collection of Industrial and Marine Bacteria Ltd., Aberdeen.

Activated sludge was obtained from three wastewater treatment sites: Buckland, Newton Abbot, Devon, (approx. 80% domestic, 20% industrial sewage); East Hyde, Bedfordshire (approx. 65% domestic, 35% industrial), and Bishops Stortford, Essex (approx. 75% domestic, 25% industrial). After washing, the final supernatant was decanted and the remaining activated sludge was fed with 50 ml OECD synthetic sewage[29] feed per litre of sludge and aerated overnight at 20 °C.

2.2 Construction of the Electrochemical Cell

2.2.1 Electrodes. Screen printed carbon electrodes on polycarbonate substrates were developed with Danielson Ltd., Aylesbury, with a circular working surface of 5 mm diameter. Ag/AgCl reference electrodes were comprised of chloridised silver rods.

2.2.2 Harvesting. Cells from single species cultures were harvested six hours after inoculation and a suspension with an $O.D._{430}$ of 1.5 (0.05 was prepared by dilution of the batch culture with sterile nutrient broth. Samples (1 ml) of diluted cultures or the overnight aerated activated sludge were centrifuged at

10^4 rev. min^{-1} for 2 minutes, and the supernatant discarded. The cell pellet was resuspended in 1 ml sterile 0.85% saline, recentrifuged for 2 minutes and the supernatant again discarded. This washing procedure was repeated. Finally, the washed cell pellet was resuspended in 100 μl sterile 0.85% saline and the cell suspension used within 30 minutes.

2.2.3 Immobilisation of the biocatalyst. The bacterial biocatalyst was immobilised onto 5 mm diameter, 0.2 (m pore size Anopore membranes (Whatman International, Maidstone, Kent) as previously described.[20]

2.2.4 Monitoring cellular metabolism using mediated amperometry. The biosensor electrodes were monitored in stirred vials containing 14.9 ml solution of respiratory substrates in 0.85% saline (sodium lactate, sodium succinate and glucose each at 5 mmol l^{-1} for *E. coli* and *Ps. putida*, and 10 mmol l^{-1} for activated sludge). The working and reference electrodes were set into the vial lid (Figure 1). Up to 15 electrode assemblies were monitored using a purpose built 15 channel potentiostat. Labtech Notebook (Laboratory Technology Corporation) provided the software interface between the operator and biosensor instrumentation, and data were displayed as current against time plots. Biosensor response was recorded every four seconds.

2.3 Biocatalyst Preservation

Harvested cells were resuspended in sterile 5% meso-inositol in nutrient broth. The suspensions were allowed to stand at room temperature for 30 minutes after which they were transferred to freeze-drying vials and cooled in a controlled manner. Finally, the vials were dried overnight in a vacuum dryer. Freeze-dried cells were resuscitated when required by resuspending in 0.85% saline containing appropriate substrates.

2.4 Chemicals

Nutrient agar and nutrient broth were supplied by LabM. NaCl and the substrates were of analytical reagent grade. 3,5-Dichlorophenol (3,5-DCP), an OECD reference toxicant,[29] and a number of site and process effluents selected by ZENECA Ltd., were used for the toxicity testing programme. 3,5-DCP solutions were made up in deionised water 24 hours prior to testing. The effluents were tested as soon as practicable in order to avoid the possibility of any temporal changes in their toxicity. All test solutions and effluents were adjusted to a pH of 7.0 ± 0.5.

2.4.1 Mediators. Potassium hexacyanoferrate (III) was used as a redox mediator to monitor bacterial metabolism.

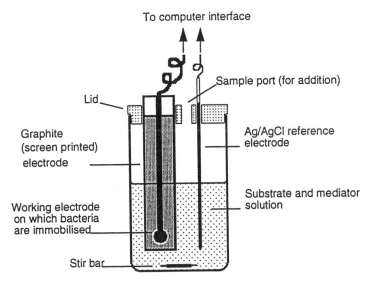

Figure 1 *Biosensor sample analysis apparatus featuring a cell loaded screen-printed electrode immersed, alongside a Ag/AgCl reference electrode, in a constantly stirred substrate and mediator solution*

2.5 Biosensor Protocol

Following a biosensor stabilisation period of 5 minutes, 100 µl of concentrated redox mediator were added to the substrate cocktail by injection via the sample port in the vial lid to give the required final concentrations of 5 mg l^{-1}. The response of the electrodes in the substrate and mediator mixture was monitored for at least 5 minutes to allow stabilisation of the electrode response, after which toxicant addition was made via the sample port. For controls, the same volume of 0.85% saline was added. Biosensor response was monitored for a further 30 minutes. Toxicity was measured by determining the degree of inhibition of the biosensor signal after a 30 minute exposure to toxicant challenge.

An empirical model in which a hyperbolic transfer function was used to relate the biosensor response to the toxicant exposure was applied in order to assess inhibition. The model used the whole of the available data, or a user defined subset of it, and allowed the long term inhibition to be predicted from the short term response. All toxicity assessments by the biosensors were calculated using this technique.

2.6 Assessment of Toxicity by Established Microbiotests and Standard Ecotoxicological Assays

A number of methods were used for comparative purposes. The assays used were; Microtox™,[10] *Tisbe battagliai* acute toxicity test (BEL SOP AQ 112),

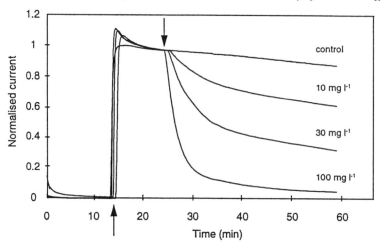

Figure 2 *Effect of 3,5-DCP on the metabolic status of a 6 h culture of E. coli immobilised on a screen printed, carbon graphite electrode, biosensor. The first arrow marks the point of addition of 5 mM potassium hexacyanoferrate (III) mediator. The second arrow marks the point of the addition of various concentrations of 3,5-DCP (labelled above traces).*

the Brixham activated sludge respiration inhibition test (BEL SOP BI 27), algal (*Skeletonema costatum*) inhibition of biomass production test (BEL SOP BA 226), acute toxicity to *Daphnia magna* static test (BEL BA 272) and the rainbow trout (*Oncorhynchus mykiss*) acute toxicity test (BEL SOP BA 75).

3 RESULTS AND DISCUSSION

3.1 Monitoring Respiration

Figure 2 shows the typical response of a biosensor to the addition of mediator following a stabilisation period in a medium containing the respiratory substrates and when challenged with a range of concentrations of 3,5-DCP.

Results obtained with fresh and freeze-dried activated sludge were very similar in both respiration levels and sensitivity to toxicants. Maintenance of response following preservation suggests that only the bacterial component was contributing to the biosensor signal. This was confirmed by microscopic observation that only bacteria survived freeze drying.

3.2 Toxicity Tests

Fifty site and process effluents were tested. Figure 3 shows typical results obtained with biosensors challenged with various effluents. Normalisation of the biosensor responses, at the point of toxicant or control addition, allowed differences between individual sensors to be taken into account, without compromising the relative responses to toxic challenge. A small gradual

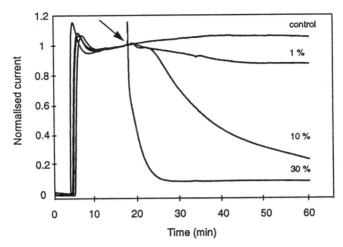

Figure 3 *Typical effect of a toxic process effluent on the metabolic status of* Ps. putida *immobilised on a screen printed, carbon graphite electrode, biosensor. The arrow marks the point of addition of various concentrations of the effluent (labelled above traces). Note the sharp peak due to sample addition.*

decline in control responses is to be expected, as the interrogation process reduces the ability of the cell to generate the energy needed to drive biosynthetic processes.[30] All toxicity assessments were made relative to the control and were calculated using the empirical model.

Biosensor response to toxic challenge most commonly takes the form of a suppression of metabolic activity related to the concentration of the toxicant. Stimulation of metabolism is seen when (i) the effluent acts as a metabolic uncoupler, (ii) the effluent is treated as a food source, or (iii) the toxicant concentration is very low allowing the biocatalyst to counter the toxic challenge by increasing its metabolic rate. Whilst the latter may result in an increased biosensor response that is sustained for a long period, as seen with activated sludge exposed to 1 mg l^{-1} 3,5-DCP (Figure 4), the apparent stimulatory effect of metabolic uncouplers is rarely sustained and is quickly replaced with a marked inhibition relative to the controls. Sharp signal spikes (Figures 3 and 4) are thought to be artifacts associated with sample introduction.

In order to allow comparison of the results obtained from biosensor tests with conventional tests, EC50 values were calculated for each of the fifty effluents using the established bioassays and each of the biosensors. Biosensor EC50 values were defined as the concentration of toxicant/effluent that would cause a 50% reduction in signal relative to the control, using 30 minute exposure data for the mathematical model to predict the ultimate inhibition. As an example, the dose response curve of the *E. coli* biosensor for increasing concentrations of 3,5-DCP (Figure 5) shows an estimated EC50 of 8.8 mg l^{-1}. The EC50 values that were generated demonstrated the marked differences in sensitivity between single species and consortia, and between different test species and protocols, especially when exposed to complex effluents.

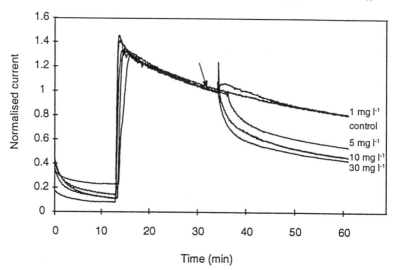

Figure 4 *Effect of 3,5-DCP on the metabolic status of activated sludge immobilised on a screen printed, carbon graphite electrode, biosensor. The arrow marks the point of addition of varying concentrations of toxicant (concentrations shown alongside traces). Note the sharp peak due to sample addition and the sustained increase in metabolic level after the addition of 1 mg l^{-1} 3,5-DCP.*

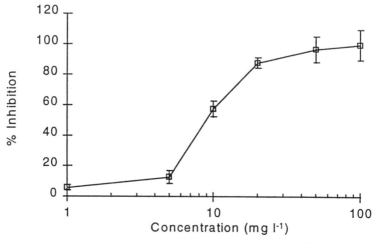

Figure 5 *The inhibition of magnitude of an E. coli based biosensor signal after the exposure to varying concentrations of 3,5-DCP. An EC50 value of 8.8 mg l^{-1} is observed. The experiment was carried out five times and the bars represent standard deviation (n-1).*

Preliminary results show that the toxicity data generated using the biosensors correlated significantly with those data obtained from the appropriate conventional tests. Preliminary correlation coefficients were calculated between; *E. coli* biosensor data and Microtox™ data, *E. coli* biosensor data and trout acute toxicity data, Microtox™ data and trout acute toxicity data, and

Table 1 *EC50 values (mg l⁻¹) for activated sludge based biosensors after 30 minute exposure to 3,5-dichlorophenol*

Source of activated sludge	% Industrial sewage	Concentration of 3,5-DCP (mg l⁻¹)
Newton Abbot, Devon	20	12.5
Bishops Stortford, Essex	25	18.5
East Hyde, Bedfordshire	35	> 100

between the data from the different biosensors. Results showed that the correlations between the *E. coli* biosensor and Microtox™ and between Microtox™ and fish were significant at the 0.1% level. Correlation between *E. coli* biosensor data and trout data was also significant at the 0.1% level. Data from the *E. coli* and *Ps. putida* biosensors compared significantly at the 0.1% level, whilst data were less comparable, although significant at the 5% level, between the *E. coli* and the activated sludge based biosensors. This is not surprising due to the considerable differences between a pure bacterial species and a complex consortium of organisms.

That the *E. coli* biosensor shows a significant correlation with Microtox™ is useful since the Microtox™ assay, although fast and reproducible, can only assess directly 50% effluent, and that it is troubled by coloured and turbid effluents.

It is interesting to note that activated sludges from sites with different wastewater influent profiles show marked differences in sensitivity. The sensitivity to 3,5-DCP of biosensors incorporating activated sludge from one of three sources decreased as the level of sewage of industrial origin in the influent increased (Table 1). Previous studies at Luton (unpublished) have shown the activated sludge from East Hyde to have remarkable levels of tolerance to organics and heavy metals (e.g. $HgCl_2$ gave an EC50 > 50 mg l⁻¹).

4 CONCLUSIONS

Both single species and mixed species consortia can be employed as biosensor biocatalysts for rapid toxicity assessment of industrial effluents. The generic methodology employed in mediated amperometry has proved to be very flexible. Mediator selection and operating protocols can be designed to favour a chosen biocatalyst, allowing species selection on the grounds of environmental relevance and sensitivity profiles rather than criteria set by the biosensor configuration. In addition, the method of bacterial immobilisation can be varied to assist the use of less robust biocatalysts. A range of bacterial immobilisation techniques have been used in this laboratory, with the filter sandwich method being chosen for the present study. In developing rapid toxicity tests, the final methodology is inevitably a compromise between optimum sensitivity and assay time. In determining EC50 values after 30 minutes exposure, we have attempted to provide a test which is sufficiently rapid to allow frequent testing, whilst providing an exposure period sufficient to allow characteristic toxic response profiles of the test species to be obtained.

A comparison of sensitivity of pure cultures and mixed consortia to both single chemicals and effluents showed interesting trends. In general, single species show greater sensitivity to toxic challenge than consortia, although the variation in response to a single chemical toxicant is less than to complex effluent mixtures.

The most sensitive species approach to biocatalyst selection is not always appropriate, and the ultimate application of the data often dictates the nature of the test organism to be used. Thus, in wastewater treatment plant protection, the most appropriate biological test material may be that which most closely resembles the plant's indigenous microbial population. Test species selection is less critical in situations where monitoring procedures are directed simply at determining changes in toxicity over time, and not at protection of a particular environment or impact assessment.

The ability to monitor selected test species of bacteria using mediated amperometry allows the exploitation of bacterial bioassays for ecotoxicological testing of effluents in biosensor configurations. There are many advantages of this approach over the conventional bioassay methods, in particular the short test period, ease of use, and the use of appropriate test species. Such systems are increasingly needed to enable a cost effective implementation of the toxicity based consent procedures for industrial effluent.

References

1. G. A. Bagdasar'yan and K. V. Genlatulin, *Gigiena I Sanltarlya*, 1981, **0**, 11.
2. D. Berkowitz, *Vet. Hum.Toxicol.*, 1979, **21**, 422.
3. G. Bitton and B. J. Dutka, 'Toxicity testing using microorganisms vol. 1', CRC Press, Boca Raton, Fla., 1986, p.32.
4. J. H. Bowdre and N. R. Krieg, *Vpi-Vwrrc*. 1974, **69**, 61.
5. J. Cairns, *Hydrobiologia*, 1989, **188**, Xiii.
6. J. W. Gillett, M. D. Knittel, E. Jolma and R. Coulombe, *Environ. Tox. and Chem.*, 1983, **2**, 185.
7. P. Lenz, R. Sussmuth and E. Selbel, *Tox. ass. int. J.*, 1989, **4**, 43.
8. A. M. Stern, *Amer. Soc. Micro.*, 1980, 361.
9. G. W. Suter, *Environ. Int.*, 1983, **9**, 157.
10. Beckman Instruments Inc., 'Microtox™ System Operating Manual', 1982, Carlsbad, CA.
11. A. A. Bulich, 'Toxicity testing using microorganisms vol 1', CRC Press, Boca Raton, Fla., 1986, p.57.
12. S. Lee, M. Suzuki, I. Tamiya and I. Karube, *Analytica Chimica acta*, 1991, **244**, 201.
13. A. A. Quereshi, R. N. Coleman and J. H. Paran, 'Toxicity Screening procedures using Bacterial Systems' 1984, p. 1.
14. J. M. Ribo and K. L. E. Kaiser, *Chemosphere*, 1983, **12** (11/12), 1421.
15. J. M. Ribo, J. E. Yang and P. M. Huang, *Hydrobiologia*, 1989, **188/9**, 155.

16 M. H. Schiewe, E. G. Hawk, D. Actor and M. M. Krahn, *Can. J. fish. Aquat. Sci.*, 1985, **42**, 1244.
17 G. S. A. B. Stewart, *Lett. App. Micro.*, 1990, **10**, 1.
18 T. K. Van Dyk, W. R. Majarian, K. B. Konstandnov, R. M. Young, P. S. Dhurjah and R. A. LaRossa, *Appl. Environ. Micro.*, 1994, **60** (5), 1414.
19 A. L. Atkinson and D. M. Rawson, 'Ecotoxicology of soil organisms', CRC Press, Inc., 1994, p 113.
20 N. J. Richardson, S. Gardner, and D. M. Rawson, *J. Appl. Bact.*, 1991, **70**, 422.
21 N. J. Richardson, Ph.D. thesis, Cranfield Institute of Technology, 1991.
22 D. M. Rawson, A. J. Willmer and M. F. Cardosi, Toxic. Ass. Int. Quart. Rev., 1987, **2**, 325.
23 D. M. Rawson, A. J. Willmer and A. P. F. Turner, *Biosensors*, 1989, **4**, 299.
24 W. Bains, *Int. Environ. Tech.*, 1994, Aug.35.
25 W. C. Gaisford, N. J. Richardson, B. G. D. Haggett and D. M. Rawson, *Biochem. Soc. Trans.*, 1991, **19**, 15.
26 F. M. Van Hoof, E. G. De Jonghe, M. G. Briers, P. D. Hansen, H. J. Pluta, D. M. Rawson and A. J. Wilmer, *Environ. Toxicol. Wat. Qual. Int. J.*, 1992, **7**, 19.
27 I. Karube and S. Suzuki, *Proc. in Biotech.*, 1983, **83**, 625.
28 I. Karube, K. Huamoto, M. Kawarai and K. Sode, *Membrane*, 1989, **14**, 311.
29 OECD technical report, 1976, June.
30 P. F. Turner, G. Ramsey and I. J. Higgins, *Biochem. Soc. Trans.*, 1983, **11**, 445.

Application of Routine Ecotoxicological Screening Methods for Assessing Suitability of Trade Effluents for Biological Treatment

K. Wadhia, A. Colley and K. C. Thompson

ALcontrol UK*, Yorkshire Environmental, Templeborough House, Rotherham S60 1BZ, UK

ABSTRACT

For some trade effluents, conventional chemical-based consents have been found to be inadequate for ensuring that the effluent has no significant effect on the receiving sewage works. Many of these effluents are complex and variable and can have a significant effect on sewage works performance. Consequently it has become necessary to develop laboratory tests that can give a much more reliable indication of the potential effect on an effluent discharge on a sewage works.

These tests must exhibit performance data (repeatability and reproducibility) that are suitable for a consent parameter. The development of these robust tests and associated quality system suitable for use in a routine laboratory is described.

Some examples of the comparison between the chemical and ecotoxicity measurements on selected effluents are given. A number of anomalous ecotoxicity results from particular effluents have been observed.

The potential of enhanced chemiluminescence tests (horseradish peroxidase-enhancer-luminol) as a rapid screening technique for monitoring effluents has been evaluated. The results from this simple type of test have been compared with the more elaborate laboratory-based tests for a number of effluents. A summary of the results obtained is presented. The advantages and disadvantages of these rapid tests are highlighted.

1 INTRODUCTION

For some industrial effluents, conventional chemical-based consents have been found to be inadequate for ensuring that the effluent has no significant effect

* Formerly LabServices

on the receiving sewage works. Many of these effluents are complex and variable with significant effect on sewage works performance,[1] highlighting the need for regulation of toxic trade wastes to sewer.[2] Consequently, it has become necessary to develop laboratory based ecotoxicity methods for regulatory purposes that can give a more reliable indication of the potential effect of an effluent discharge on a sewage works. In addition to general screening methods, ecotoxicity methods are required for monitoring the specific biological processes in sewage treatment works (STWs).

In order to assess whether particular effluents reduce optimal performance of the biological processes at STWs, on-site and laboratory tests can be employed. Ecotoxicity methods are required for these specialised groups of prokaryotic and eukaryotic organisms which can have different cellular physiology and sensitivity to toxicants compared with a standard monoculture, such as *Photobacterium phosphoreum* (syn *Vibrio fischeri*) as used in the Microtox^{TM} test.

The selected ecotoxicity methods must therefore be sensitive, robust and relevant to specific biological processes. The methods will routinely be employed on samples from a wide range of industrial processes, *e.g.* food processing, chemical manufacture, waste disposal, wool treatment, textile manufacture and dyeing, *etc*. These tests must generate performance data that are suitable for consent parameters to protect the STWs. The development of these robust tests and associated quality system suitable for use in a routine laboratory to external accreditation standards will be described.

The potential of enhanced chemiluminescence tests (horseradish peroxidase-enhancer-luminol) as a rapid screening technique for monitoring effluents has been evaluated. The results from this simple type of test have been compared with the more elaborate laboratory-based tests for a number of effluents. Many other ecotoxicity methods could also be applied to effluent testing including, *Daphnia magna* neonate immobilisation and *Lemna minor* growth inhibition.

2 FIELD AND LABORATORY SCREENING METHODS

Whilst field tests are currently being developed, such as Microtox-OS and biosensors for specific bioassays, they were not readily available when initial development of protocols was commenced in the 1980s by Yorkshire Water Services (YWS).

2.1 Enhanced Chemiluminescence Kits

One recent development for on-site and laboratory testing is enhanced chemiluminescence, based on a steady state of chemiluminescence as luminol reacts with an oxidant to produce light. The enzyme horseradish peroxidase (HRP) catalyses the reaction between the oxidant and a chemical enhancer, to

Table 1 Comparison of Ecotoxicity Methods for Industrial Effluents and 3,5-DCP

Toxicity Method	Industrial Effluents and EC50 (x dilution)					3,5-DCP EC50 (mg l^{-1})
	1 A	1 B	2	3	4	
Aquanox^{TM}	210	234	5677	153	59	112
Eclox^{TM}	1285	N/A	4160	N/A	N/A	112
Nitrification Inhibition	10.9	N/A	369	26	17	1.7
Microtox^{TM} (5 minutes)	10.7	19	243	40	33	4.3
Respiration Inhibition	N/A	1.5	34	12	6.6	14.4

form an enhancer radical. This reacts with the luminol to form a luminol radical which emits light, detected in a luminometer. Light output can be reduced by chemicals which scavenge free radicals, for example anti-oxidants (*e.g.* ascorbic acid) or inhibitors of the HRP enzyme (*e.g.* cyanides, amines, phenols and heavy metals). The reduction in light output is usually proportional to the concentration of contaminant present. The technology has been licensed to companies which supply monitoring systems, for example Aztec Ltd for the Eclox system and Randox Laboratories Ltd for the Aquanox system.

It should be noted that the two systems described have undergone some modifications since the assessments carried out in 1995. These chemiluminescence toxicity methods are considered to be semi-quantitative screening tests. They are relatively inexpensive and can be used for the rapid detection of pollutants or tracing significant changes in industrial process or natural water quality. They are not primarily intended and were not considered suitable for monitoring of trade effluents on a regulatory basis. The majority of work involved assessment of sensitivity to a range of individual toxicants and components expected in effluents. Some examples of comparison of enhanced chemiluminescence kits with other ecotoxicity methods are given in Table 1.

2.2 *Photobacterium phosphoreum* – Microtox

The Microtox^{TM} system has been used by this laboratory for several years to assess the toxicity of effluents. The large scale production and ability to freeze-dry *Photobacterium phosphoreum* both contribute to the good repeatability between batches when using this test.

To determine within batch variation, two operators simultaneously carried out a series of 20 EC50 determinations for 3,5-DCP (Table 2). The within

Table 2 *Performance Data for Microtox (EC50 5 mins) with 3,5-dichlorophenol*

	EC50 (mg l^{-1})	SE (mg l^{-1})	CV (%)	n
Within batch - A	4.17	0.028	0.67	20
Within batch - B	4.28	0.019	0.44	20
Between batch	4.34	0.039	3.98	20

Within batch: Operator and machine A; operator and machine B.
Between batch: 20 batches of reagents.
SE = Standard Error
CV = Coefficient of Variation
n = Number of replicates

batch coefficient of variation (CV) of less than 1% indicates good repeatability, with little significant difference between operators or machines. The 'between batch' data includes different batches of all reagents and toxicants involved. The between batch variation of 3.98% CV also indicates satisfactory repeatability, when compared with routine chemical methods with up to 20% CV and some ecotoxicity methods quoted at 20-80% CV.

3 LABORATORY BASED BIOASSAYS FOR SPECIFIC PROCESSES

In addition to general screening methods, ecotoxicity methods that are relevant to specific biological processes are required. The methods to investigate inhibition to nitrification[3,4,5] and respiration[6,7] are based upon standard protocols.

Since publication of the UK Standing Committee of Analysts (SCA) protocol to determine inhibition to nitrification,[3] pure cultures of the nitrifying bacteria have become commercially available (with production to ISO 9002 standards). This advance has been incorporated into the method used by ALcontrol UK with the aim of improving repeatability and reproducibility, with the assurance of supplies for daily testing throughout the year.

Assessment of the degradation of organic material by active sludge in a sewage treatment works can be carried by respirometry techniques. Performance data have been obtained using Merit respirometers, now used for routine testing.

3.1 Determination of Inhibition to Nitrification

3.1.1 The Role of Nitrifying Bacteria in Effluent Treatment. Oxidation of ammonia is primarily carried out by the autotrophic nitrifying bacteria. *Nitrosomonas* oxidises ammonia to nitrite, which is the rate limiting step for the overall process.[8] The metabolite nitrite is a substrate for *Nitrobacter*, which is oxidised to nitrate. The activated sludge tanks in sewage treatment works contain natural populations of these two organisms.

The physiological specialisation of these organisms is one factor which makes this biological process more susceptible to toxicity present in effluents. The low growth rates of nitrifying bacteria can also result in slow recovery rates following a toxic incident. Consequently, it is important to assay for potential disturbance to the biological nitrification process by industrial effluents. Instead of maintaining activated sludge containing a low proportion of nitrifying bacteria, pure cultures of nitrifying bacteria are employed in this method.

3.1.2 Method for Determination of Nitrification Inhibition. The sample was buffered with $KHCO_3$ to maintain optimum pH, dosed (or diluted if necessary) to provide a standard concentration of ammoniacal nitrogen and a calibrated dose of nitrifying bacteria (International Biochemicals Ltd, Dublin) added. The cultures were aerated at 20 °C to provide approximately 80% nitrification in the non-toxic controls over the four hour test period. Measurement of the initial and final ammonia and/or TON concentrations allows for calculation of the nitrification rate over the test period.

3.1.3 Performance Data for Nitrification Inhibition. Within and between batch performance data have been obtained for 3,5-DCP. Additional chemical toxicants tested were zinc sulphate, copper sulphate and phenol. Within batch variation is also reported for a number of industrial effluents.

The nitrifying bacteria were as sensitive to 3,5-DCP (Table 3) as *Photobacterium phosphoreum* (shown previously in Table 2). The between batch data were collected from 20 different batches of nitrifying bacteria and 3-20 different batches of all other reagents and 3,5-DCP. The nitrification inhibition assay showed similar sensitivity to phenol, zinc sulphate and copper sulphate compared with 3,5-DCP, although with slightly higher within batch variation than for the 3,5-DCP.

When testing effluents, the EC50 value is expressed as the final sample dilution, as for the US EPA 'toxicity unit'. For the industrial effluents listed in Table 4, the EC50 values ranged from x 1.1 to x 369 dilution, with coefficients of variation similar to those for individual chemical toxicants. The standard error and %CV quoted for effluent 3 are for four replicate measurements at x 1.1 sample dilution.

3.2 Determination of Inhibition to Respiration Using Activated Sludge

In addition to an assay specific to the nitrification of ammonia, oxidation of the total carbonaceous component of effluents needs to be monitored. The range and complexity of prokaryotic and eukaryotic organisms involved in activated sludge precludes the use of pre-prepared cultures. Activated sludge is

Table 3 *Performance Data for Nitrification Inhibition with Chemical Toxicants*

Within batch testing:	EC50 (mg l^{-1})	SE (mg l^{-1})	CV (%)	n
Phenol	1.26	0.19	33.4	5
ZnSO$_4$	2.29	0.16	18.6	7
CuSO$_4$ (as Cu)	2.24	0.21	22.9	10
Within batch: 3,5-dichlorophenol	1.7	0.09	15.4	10
Between batch: 3,5-dichlorophenol	2.4	0.16	29.9	20

Between batch: 20 batches of reagents and nitrifying bacteria.
SE = Standard Error
CV = Coefficient of Variation
n = Number of replicates

Table 4 *Performance Data for Nitrification Inhibition for Industrial Effluents*

Industrial effluent	EC50 (x diln)	SE (x diln)	CV (%)	n
1	17.6	0.6	12.1	11
2	369	2.5	1.2	6
3*	@ x 1.1: 54.7%	2.7	4.9	4

* Low nitrification inhibition for Effluent 3; results quoted are from one dilution tested.

taken from STWs treating predominantly domestic waste, to ensure no acclimatisation to toxic compounds. This is maintained in the laboratory for use in the endogenous phase during testing.

3.2.1 Range of Methods Available for Determination of Respiration Inhibition. In addition to the routine protocol, a sample can be tested simultaneously using unacclimatised and acclimatised activated sludge. Oxygen uptake rates can also be determined for samples in the absence of an added nutrient source (synthetic sewage). Many established protocols are available for short term respirometry (*e.g.* 5 to 240 minutes) through to biodegradation (28 days or longer).

3.2.2 Protocol for 3 Hour Test using Merit^{TM} Respirometer. Automated, processor controlled manostatic respirometers were used (Merit^{TM} Respirometer, Terra Nova, York). The method employed was based upon the 1982 HMSO publication 'Methods for Assessing the Treatability of Chemicals and Industrial Waste Waters and their Toxicity to Sewage Treatment Processes.'[6] This method has been updated, using the Merit respirometer and is awaiting publication as a new SCA method.[7]

Activated sludge is collected from a STW treating domestic effluent, maintained in the laboratory and used in the endogenous phase. Activated sludge prepared to 6000 mg l^{-1} suspended solids is mixed with a standard synthetic sewage and the test sample, non-toxic control or standard toxicant. The standard quantities of activated sludge and synthetic sewage provide a suitable respiration rate and ratio of nutrients to biomass (F/M). The test material may be water, sewage, trade effluent, process effluent or a standard chemical toxicant.

The test vessels are sealed into the respirometer and oxygen uptake determined from the amount of oxygen generated electrolytically to maintain a constant gas pressure in each test vessel. The respirometer is operated at 20 °C, with 15 minutes equilibration followed by a three hour test period. As oxygen and organic compounds are respired to carbon dioxide, this gas is absorbed by caustic material causing a reduction in the gas pressure within the sealed system. This triggers a pressure senstive switch, supplying a constant current to an electrolytic cell, generating oxygen to balance the gas pressure. The quantity of oxygen generated is recorded at pre-determined time intervals; hourly in this method. All testing is carried out in duplicate. The rate and mass of oxygen uptake by activated sludge is monitored for each test mixture enclosed in the respirometer.

The control respiration rate is the mean oxygen uptake rate for the 2 to 3 hour period for activated sludge with synthetic sewage in the presence of non-toxic sewage or water. The percentage inhibition to respiration due to samples or standard toxicants is calculated compared with the control respiration rate.

3.2.3 Determination of Suitability of Activated Sludge. Activated sludge was collected from three STWs on a number of occasions and the EC50 for 3,5-DCP determined. The results showed similar sensitivities to 3,5-DCP, with the EC50 ranging from 8.1 to 9.2 mg l^{-1}. In contrast, the EC50 for 3,5-DCP using activated sludge from works receiving a mixture of domestic and industrial wastes exceeded 20 mg l^{-1}.

3.2.4 Use of the Shewhart Control System for the Respiration Inhibition Method. The QC data for 3,5-DCP from January 1995 to March 1996 are shown in Figure 1. The actual EC50 value is displayed with the warning ($\pm 2SD$) and failure limits ($\pm 3SD$) as a Shewhart chart. For the activated sludge used, the mean EC50 was 10.39 ± 2.67 mg l^{-1} (SD), n=163. Investigation of 'warning' or 'failure' level results have been traced to problems with operator, machine, reagents or the biological test components.

Activated sludge from a range of STWs has provided data within these predetermined quality control limits for both control respiration rates and sensitivity to toxicants. Activated sludge from any domestic sewage works must be accepted by these QC procedures prior to routine use.

3.2.5 Performance Data for Respiration Inhibition. Testing was performed to assess the effect on respiration inhibition with a range of chemical toxicants (Table 5). Of the chemicals tested, 3,5-DCP and 3,4-dichloroaniline (3,4-DCA) were most inhibitory to respiration. Zinc sulphate and copper sulphate were less toxic than the two chlorinated organic compounds tested. The respiration process was least sensitive to phenol as a toxicant.

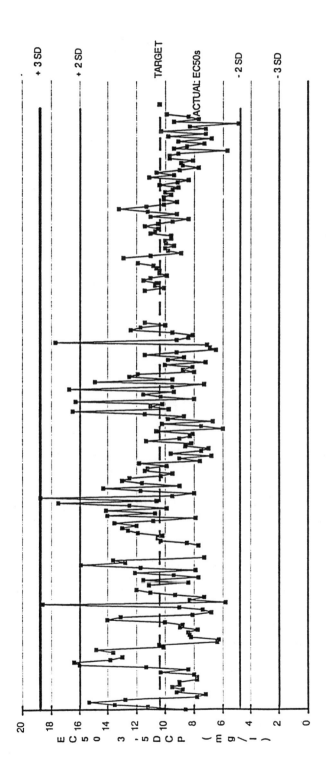

Figure 1 *Respiration Inhibition control results for 3,5-dichlorophenol*

Table 5 *Performance Data for Respiration Inhibition for Chemical Toxicants*

Within batch testing:	EC50 (mg l^{-1})	SE (mg l^{-1})	CV (%)	n
Phenol	670	32.1	15.2	10
3,4-DCA	57	1.4	15.2	10
ZnSO$_4$	112	3.8	10.8	10
CuSO$_4$	119	1.0	6.9	10
Within batch: 3,5-dichlorophenol	14.4	0.46	10.1	10
Between batch: 3,5-dichlorophenol	9.8	0.41	18.7	20

Between batch: 20 batches of activated sssludge, simpled from one STW.

Table 6 *Performance Data for Respiration Inhibition for Industrial Effluents*

Industrial effluent	EC50 (x diln)	SE (x diln)	CV (%)	n
1	10.64	0.22	6.57	10
2*	@ x 1.54: 35.3%	0.80	7.20	10
3**	2.01	0.11	15.2	8
4	.92	0.08	8.15	10

Low toxicity by respiration inhibition:
  * results quoted from most toxic dilution tested.
  ** two additional sets of data gave EC50 < x 1.54 dilution.

Several industrial effluents were tested (Table 6), with 10 replicate EC50 values obtained for three samples with CVs from 6.6% to 15.2%. Industrial effluent 2 tested at a final dilution x 1.54 resulted in less than 50% respiration inhibition. The mean value of these ten replicates was 35.3% 0.8% (standard error). Two replicate dilutions of industrial effluent 3 resulted in inhibition of less than 50%; the data were not extrapolated to produce an EC50 value.

4 COMPARISON OF ECOTOXICITY DATA

The data presented in Tables 1 to 6 show that the response to individual chemicals or effluents can vary by orders of magnitudes between different ecotoxicity tests. Experience of routine laboratory workload has demonstrated that on some occasions, low toxicity indicated by one method can not be correlated with low inhibition to the nitrification or respiration processes of STWs.

5 CONCLUSIONS

No single ecotoxicity method will determine the overall toxicity of an individual effluent. Screening methods can be valuable for rapid screening or repeat monitoring of individual effluents. A suite of ecotoxicity methods is required to determine the potential toxicity of industrial effluents to the biological processes in sewage treatment. The methods described for nitrification inhibition and respiration inhibition have been developed to provide robust, repeatable tests and quality control protocols.

Routine monitoring with a suite of ecotoxicity tests has allowed for detection of effluent toxicity to individual stages of sewage treatment, *e.g.* low toxicity by the Microtox test, but complete inhibition of activated sludge in the respiration test, or low respiration inhibition but extremely high toxicity to the nitrification process.

The selection of particular tests must be balanced against the information required for an individual sample. The reasons for sampling can be as varied as the aquatic samples to be tested, *e.g.* introduction of a new industrial process, routine monitoring of effluents, additional information to back up the findings of chemical analyses, emergency sampling following an effluent spillage or tracing the cause of poor performance for an industrial effluent plant or sewage works.

ACKNOWLEDGEMENT

The commitment and expertise of staff and management of ALcontrol UK and YWS Ltd are acknowledged. It is through their endeavours that ALcontrol UK have developed Ecotoxicity methods for UKAS accreditation.

DISCLAIMER

The opinions expressed in this chapter are those of the authors and do not necessarily represent those of ALcontrol UK or Yorkshire Water PLC.

References

1. Fearnside, D and Hiley, P, The role of Microtox in the detection and control of toxic trade effluents and spillages. *In:* Ecotoxicology Monitoring, Ed M. Richardson, Ch 23, 319-332, VCG, Cambridge, 1993.
2. Public Health (Drainage of Trade Premises) Act, 1936, HMSO, London.
3. The Assessment of the Nitrifying Ability of Activated Sludge. Methods for the Examination of Waters and Associated Materials. HMSO, 1983.
4. Biotox NH_3, International Biochemicals, Dublin, 1990.
5. Biotox NO_2, International Biochemicals, Dublin, 1992.
6. Methods for Assessing the Treatability of Chemicals and Industrial Waste Waters and their Toxicity to Sewage Treatment Processes. Methods for the Examination of Waters and Associated Materials. HMSO, 1982.
7. Determination of the Inhibitory Effects of Chemical and Industrial Waste Waters on the Respiration of Activated Sludge using a Manostatic Respirometer. Methods for the Examination of Waters and Associated Materials. Draft for HMSO, 1996.
8. Young, J C. Chemical Methods for Nitrification Control, J. Water Poll. Control Fed., **45**(4), 637-646.

Eclox™: A Rapid Screening Toxicity Test

E. Hayes and M. Smith

Severn Trent Water Ltd, Process Technology, Finham STW, St. Martins Road, Coventry CV3 6PR, UK

ABSTRACT

Toxicity is the property of a substance or mixture of substances which have a harmful effect on living organisms. A survey carried out[8] found that only two of the ten water utilities regularly used some form of toxicity tests. As the nature and diversity of chemicals discharged into the aquatic environment continues to increase, it is not feasible to monitor each chemical individually. In addition chemicals may act synergistically or antagonistically in river systems and these effects cannot be determined by effluent chemistry alone. There is a need therefore to examine the impact of effluents as a whole on living organisms.

Traditionally, toxicity tests utilise a range of whole organism assays to determine acute or chronic toxicity,[9] however, these tend to be expensive and time consuming. It is important therefore to develop rapid toxicity tests which can give an immediate indication of the quality of the sample tested.

The Aztec Eclox™ monitor uses an enhanced chemiluminescent reaction to determine the toxicity of a sample. It is a free radical reaction of the oxidation of luminol in the presence of the horseradish peroxidase enzyme using p-iodophenol as an enhancer and to stabilise the reaction. Any substance that inhibits the action of the enzyme causes a reduction in light emission which is measured on a portable luminometer and is displayed as a characteristic curve. As part of ongoing trials with this unit, surveys were undertaken to examine the reduction in toxicity through different sewage treatment processes on a range of works treating both domestic and industrial effluents. Initially, four works were evaluated:

1) a conventional biofilter plant
2) an activated sludge plant with a separate nitrification stage.
3) a small biofilter plant with reed beds,
4) a works with two streams: a biological nutrient removal plant and a conventional biofilter plant side by side.

The results show an attenuation in toxicity through each plant with increasing treatment through each plant. Primary settlement has little impact on the overall toxicity of the sewage stream but with each biological treatment stage there is a steady increase in light emission due to a reduction in toxicity.

Characteristic curves are produced for each works and deviations from the normal pattern can be further investigated with chemical analyses. The efficiency of different biological systems can be compared to assess the most effective treatment for minimising the toxic impact of an effluent on the receiving watercourse.

1 INTRODUCTION

Toxicity is the property of a substance or mixture of substances which has a harmful effect on living organisms. Toxicity based consents are already in use in the USA[1] and in Canada on a variety of effluents. However many acute and chronic toxicity tests are both expensive and time-consuming. Multicellular animals are not an ideal subject for toxicity testing due to their metabolic complexity. Single celled or isolated cells are considered more suitable subjects for toxicity testing.[2]

Current legislation in the UK requires that sewage works discharging to river meet consent conditions determined by the Environment Agency. These usually cover the conventional sanitary determinants of biochemical oxygen demand (BOD_5), suspended solids (SS) and ammoniacal nitrogen (ammN). Additional determinants such as metals and pesticides may also be consented where appropriate. Conventional chemical analyses cover $<0.1\%$ of the chemicals discharged to the aquatic environment.[3]

In order to address this issue the Environment Agency have been evaluating a range of standard and rapid tests to develop a strategy for the formulation of toxicity based consents.[4] Toxicity tests are considered to complete a triad of tests when considered alongside more conventional chemical analyses and routine biological surveys.

For a sewage works that is required to meet a toxicity based consent there will be an additional burden of monitoring to ensure compliance. The cost of this monitoring may be minimised by the use of less expensive rapid screening tests to assess compliance with the more expensive standard test only being used where a problem is identified.

There is a wide range of rapid tests available including Microtox™, Polytox™, Rodtox™, Randox™ which may be used. These work on the principle that toxins present in a sample of effluent act to inhibit the metabolism of the cells used in the assay. Microtox™ is one of the most widely used tests. It has been tested against standard toxicity tests and is considered more sensitive than many bacterial assays.[5] However, in spite of these correlations, it has also been suggested that compounds that are toxic to Microtox™ are not necessarily toxic to most aquatic organisms.[6] In addition, high quality effluents may give rise to an enhanced output.

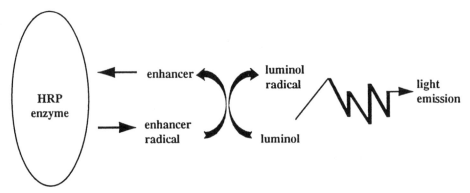

Figure 1 *Mechanics of the Reaction*

The Eclox™ is a new rapid screening test that has been developed. The system uses an enhanced chemiluminescent reaction to determine the toxicity of a sample. The reaction utilises the enzyme horseradish peroxidase (HRP) to catalyse the oxidation of luminol which results in the production of a flash of light. The addition of an enhancer stabilises the reaction to produce a prolonged emission of light which is measured using a luminometer. The presence of free radical scavengers interfere with the reaction, or toxicants in a sample interact with the enzyme, so reducing light emission. The mechanics of the reaction are shown in Figure 1.

This is a highly sensitive test requiring dilution of samples but it has the advantage that it can be used to cover the full range of samples from crude sewage to final effluent on the same test. Preliminary work indicated that this relatively simple test allows a water sample to be assessed quickly and that the test could be a useful diagnostic tool.[7]

Early work focused on the use of the instrument for the assessment of river water quality. However, with the drive towards toxicity based consents a programme was established to examine toxicity reduction through a range of sewage treatment processes.

2 METHOD

2.1 Assay Technique

The Eclox™ system uses an enhanced chemiluminescent reaction to give an empirical measure of toxicity. The test sample is diluted with distilled water to an appropriate dilution, then the enzyme, the luminol and the enhancer are added to the cuvette. The cuvette is shaken to mix the components and then inserted into the luminometer in order to measure the light output. The test is carried out over a period of up to six minutes, the output is displayed in graphical format on the luminometer as the data are generated. The information is stored in the instrument and can be downloaded to a conventional computer.

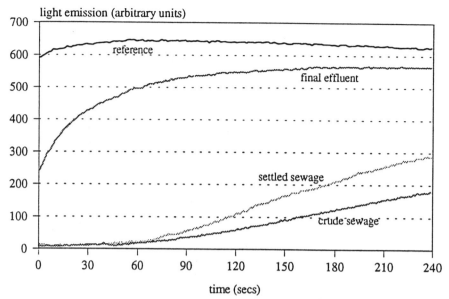

Figure 2 *Characteristic Eclox Traces for Sewages at 1:40 Dilution with Distilled Water*

There are three key methods for interpreting the data: by looking at the characteristics of the curve generated and assessing variability with time; by comparing the integration of the area beneath the reference curve with that of the sample curve: the area inhibition; or by calculation of the Eclox^{TM} Number. A typical set of curves are shown in Figure 2.

In order to compare samples that may be tested at different dilutions the Eclox^{TM} Number was determined from:

$$\text{Eclox}^{TM} \text{ Number} = \frac{\% \text{ Area Inhibition}}{100} \times \frac{1250}{\text{sample volume}} \times \text{dilution} \times 10$$

For comparison between sites an overall toxicity reduction value was calculated as a percentage based on the reduction in area inhibition at each stage of treatment.

Four works were selected for assessment as part of the instrument evaluation. A description of each of these is given in Table 1.

At each site, key sampling points were identified. These are shown on the process diagrams as a cross in Figure 3. Sites were chosen to represent a broad diversity of treatment processes. Wherever possible composite samples were taken at the points identified. All surveys were carried out during dry weather and in addition to the Eclox test all samples were submitted for routine chemical analyses.

A total of seven surveys were carried out at each site. The data for each works were collated and mean values obtained. Mean traces for works A are shown in Figure 4. Works A is a large plant treating a mixture of domestic and

Table 1 *Description of the Four Works Evaluated*

WORKS	DESCRIPTION
A	A large manned works treating both domestic and industrial waste from a large city within the West Midlands.
B	A large manned works treating both domestic and industrial waste but with a significant proportion of textile effluent.
C	A small unmanned works treating only domestic waste from a number of small villages.
D	A medium size works with a major discharge from food processing. There are two parallel treatment works on the same site.

Table 2 *Summary of Mean Chemical Results for the Four Works Studied*

WORKS	SAMPLE	BOD_5 (mg l^{-1})	COD (mg l^{-1})	AMMONIA AS N (mg l^{-1})
A	crude sewage	227	583	29
	settled sewage	144	351	34
	humus tank effluent	18	99	12
	final effluent	7	57	4
B	crude sewage	263	839	72
	settled sewage	157	403	37
	asp effluent	6	57	14
	final effluent	4	55	0.5
C	crude sewage	291	701	32
	settled sewage	255	627	41
	humus tank effluent	7	56	5
	final effluent	2	28	0.7
D	crude sewage	738	2143	49
	settled sewage	305	637	42
	humus tank effluent	14	88	3
	nutrient removal plant effluent	2	40	0.3
	final effluent	6	56	0.7

industrial waste. The data indicate that physical settlement of solids has little impact on the overall toxicity of the sample. The significant reduction in toxicity is seen as a result of biological treatment. After biological treatment the effluent undergoes tertiary treatment by sand filtration and this further reduces the toxicity of the sample before discharge. This pattern is seen at each works studied.

A typical set of curves are shown in Figure 5 for works B. These are of interest because the biological treatment comprises two stages: first an activated sludge plant removes the BOD_5, then the effluent is settled and pumped onto nitrifying filters which oxidise the ammonia to nitrate. These

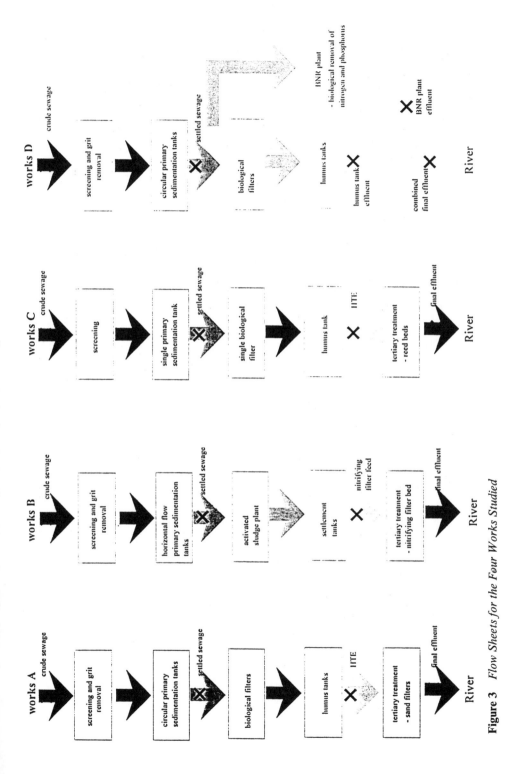

Figure 3 *Flow Sheets for the Four Works Studied*

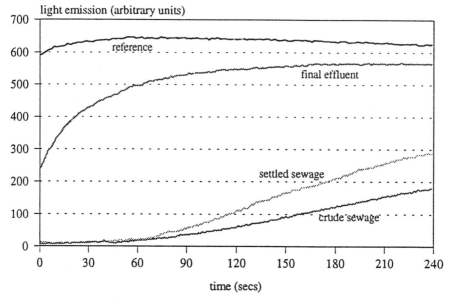

Figure 4 *Traces for Works A*

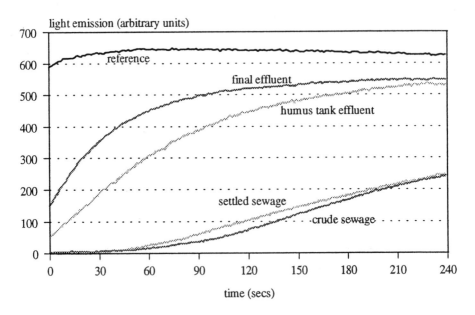

Figure 5 *Traces for Works B*

Table 3 *Results Generated from Eclox: all Samples Tested at 1 in 40 Dilution with Distilled Water*

WORKS	SAMPLE	% AREA INHIBITION	ECLOX™ NUMBER	% TOXICITY REDUCTION
A	crude sewage	89	445	–
	settled sewage	84	418	6.0
	humus tank effluent	46	228	49
	final effluent	27	134	70
B	crude sewage	94	471	–
	settled sewage	87	436	7.3
	asp effluent	45	224	53
	final effluent	32	158	67
C	crude sewage	94	470	–
	settled sewage	78	392	17
	humus tank effluent	20	102	78
	final effluent	20	98	79
D	crude sewage	97	486	–
	settled sewage	96	482	1.0
	humus tank effluent	30	152	69
	nutrient removal plant effluent	26	128	74
	final effluent	31	157	68

traces show again that primary settlement has little effect on toxicity. However, when assessing toxicity reduction at the biological stage it is the removal of BOD_5 which has the greatest effect on the Eclox output, whilst the ammonia removal results in a further improvement, this is not so significant.

Works D was chosen for study because of the ability to compare two treatment systems acting on the same sewage. These data indicate that there was no significant difference in toxicity between the two effluent samples although the effluent produced by the biological nutrient removal plant had a consistently lower BOD_5 and ammonia concentration.

The mean results for all four works are summarised in Tables 2 and 3.

3 DISCUSSION

The results indicate that each stage of the sewage treatment process results in a reduction in toxicity measured by the Eclox™. The primary settlement stage gives rise to a significant reduction in BOD_5, however this is not reflected in the Eclox™ results indicating that there is a component in the settled sewage which has sufficient toxicity to override the impact of the reduction in BOD_5. This pattern has been seen across all four works studied.

Overall, the biological treatment stage results in the greatest reduction in Eclox™ Number. This is due to the reduction in BOD_5 and the conversion of the ammoniacal nitrogen to nitrate nitrogen. The study of works B revealed that the reduction in the BOD_5 is in fact the most significant step when considering toxicity reduction, this resulted in a reduction in toxicity of 53% whereas the removal of ammonia only increased the toxicity reduction by a further 14% at this site.

On all four works, BOD_5 removal was consistently greater than 95% through the treatment process. Toxicity reduction was not so efficient with a maximum removal of 79%. This was achieved at works C which treats only domestic sewage and therefore the toxins should be present at lower concentrations than at works treating industrial effluents. The data show consistent toxicity reduction of 65–70% on the three other works, all of which treat industrial waste.

Works D has a full scale biological nutrient removal (BNR) plant running in parallel with a conventional biofilter plant. This afforded an excellent opportunity to compare toxicity reduction using two treatment systems on the same sewage. The results indicate a slightly higher reduction in toxicity on the BNR plant however this could not be considered significant. This indicates that on this sewage, either plant would achieve the same level of toxicity reduction although in terms of conventional chemical quality the BNR gave better results.

The tertiary treatment at sites A and C gave only a small improvement in the effluent quality and corresponding toxicity measure.

In order to link the response from the Eclox^{TM} with chemical analyses of the effluents, regression analyses were carried out for each works and each series of samples. When assessing all data from a particular treatment stage no simple universal correlation with the area inhibition values was determined. However, if the data from each works were considered individually then a significant correlation was seen for the three key parameters. The most significant correlation was established at each site with respect to BOD_5. In each case $r^2 > 0.85$ across all the samples tested. This suggests that Eclox^{TM} may have a role not only for toxicity monitoring but potentially also as a substitute BOD_5 test on a site for site basis.

The Eclox^{TM} is a new instrument which has shown a very promising correlation with effluent quality. In addition, the results indicate that the Eclox^{TM} can detect other quality factors which may be related to toxicity in sewage effluents or rivers independent of conventional sanitary determinants normally used to assess quality.

Eclox^{TM} shows its potential as a rapid screening test that determines quality differences. The issues that need to be resolved are whether the response of Eclox^{TM} is a demonstration of true toxicity to aquatic life and whether there is comparability between the toxicity responses of other rapid tests that are available and the responses of conventional toxicity tests.

References

1. US EPA (1991) Technical Support Document for water quality based toxics control. Office of water enforcement and permits, Washington.
2. W. Bains (1992) Sensors for a clean environment. *Biotechnology* **10** (5) 515-518.

3 J. H. Looney (1995) How toxicity reduction evaluation can help you meet your toxicity based consents. Paper presented at Solutions '95 Manchester.
4 NRA (1995) Toxicity based consents. HMSO.
5 L. Somasundarum, J. R. Coats, K. D. Racke, H. M. Stahr (1990) Application of the Microtox system to assess the toxicity of pesticides and their hydrolysis metabolites. *Bull. Environ. Contam. Toxicol.* **44** 254-259.
6 T. Brorson, I. Bjorklund, G. Svenstan and R. Lantz (1994) Comparison of two strategies for assessing ecotoxicological aspects of complex wastewater from chemical-pharmaceutical plant. *Environmental Toxicology and Chemistry* **13**(4) 543-552.
7 C. Billings, M. Lane, A. Watson, T. P. Whitehead and G. Thorpe (1993) The enhanced chemiluminescent technique. *Proc. Symposium on Analytical Sciences.* French Society of Analytical Chemistry.
8 D. W. Mackay, P. J. Holmes and C. J. Redshaw, (1989) The application of bioassay techniques to water pollution problems - the UK experience. *Hydrobiologica* **188-189**,77-86
9 J. N. Ribo and K. L. E. Kaiser (1983) Effects of selected chemicals on photoluminescent bacteria and their correlations with acute and sublethal effects on other organisms. *Chemosphere* **12** 1421-1442.

Interpretation, Relevance and Extrapolations: Can We Devise Better Ecotoxicological Tools to Assess Toxic Impacts?

M. H. Depledge

Plymouth Environmental Research Centre, University of Plymouth, Drake Circus, Plymouth, PL4 8AA, UK

ABSTRACT

The discharge of toxic wastes into aquatic environments is currently controlled by standards and limits imposed by regulators. In the main, the information upon which such standards and limits are based has been obtained under laboratory conditions using a limited number of species and test protocols. Once potential toxicity has been assessed, assumptions are made about the safety factors that must be used to prevent harmful effects on natural populations and communities. Day to day environmental management mainly involves monitoring to ensure that legal limits are not exceeded and that, where possible, guidelines are followed. Much less attention is paid to detecting the subtle, insidious changes that environmental contaminants may give rise to (changes in population distributions, genotypic make-up, community structure, ecosystem processes), perhaps over long times scales (10-20 years). In the search for better ways of assessing the risks posed by the release of specific chemicals, particular attention must be paid to improving the ecological relevance of toxicity tests, establishing a firmer scientific basis for extrapolating from laboratory test results to field situations, and to ensuring that test validation procedures are based on a sound mechanistic understanding of the responses being measured. These issues are discussed using the biomarker approach as an illustrative example of the difficulties faced in developing new and better ecotoxicological tools.

1 INTRODUCTION

It has been estimated that more than 100,000 chemicals are now in regular industrial use.[1] Choices have to be made as to how much, and over what time scales, these chemicals can be released into the environment without producing significant adverse biological effects. Day to day environmental management involves evaluating and ranking the relative toxicities of chemicals, developing guidelines and legislation for the setting of safe discharge limits, and monitoring to ensure that Environmental Quality Objectives are being met. The magnitude of the task is enormous and is further complicated by the need to achieve effective control quickly. Thus, pragmatic approaches have been developed which involve simple tests and the adoption of regulations which do not take into account the complex nature of the environment. This has resulted in ca. 30 species of animals and plants being used routinely in ecotoxicological test procedures to estimate the threats posed by chemicals to more than 30 million species world-wide! The toxicities of chemicals are usually evaluated singly, in relatively constant, favourable laboratory conditions, in contrast to the conditions pertaining in natural ecosystems where organisms face the challenge of exposure to complex chemical mixtures as environmental variables (temperature, salinity, oxygen tension, etc.) fluctuate and whilst having to proceed with such routine tasks as obtaining food, finding a mate and avoiding predators.

2 REAPPRAISAL OF THE SIGNIFICANCE AND RELEVANCE OF ECOTOXICOLOGICAL TEST PROCEDURES

The prime objective of environmental managers is to ensure that anthropogenic chemicals and radiations do not damage natural ecosystems. The science which they rely on to provide both the conceptual and practical basis for their management decisions is ecotoxicology. This is defined here as:

The scientific study of the effects of anthropogenic chemical contaminants and radiations on ecosystems and their components.

In every discipline it is important periodically to reappraise the approaches and methods in common use to see if areas for improvement can be identified. The most important ecotoxicological procedures currently used include:

(i) Quantitative Structure Activity Relationships (QSAR)[2]
(ii) Acute and chronic toxicity tests[3]
(iii) Biomonitoring[4]
(iv) Ecological monitoring[5]

It is notable that while these procedures do have practical utility in providing data upon which to base legislation and effect compliance, they do not (with the exception of ecological monitoring) yield information which contributes greatly to the assessment of ecologically-significant effects of contaminants on ecosystems. Thus, the QSAR approach allows the potential, relative toxicities of new and old chemicals to be estimated and their fate to be predicted. However, these estimates and predictions have not been properly validated *in situ*, nor do they take account of the effects that complex mixtures of chemical contaminants may have on the ecological relationships among organisms in the natural environment. The importance of this issue should not be underestimated since deaths in a population due to toxicant exposure may be of far less ecological significance than subtle chemically-induced changes in, for example, feeding activity or mate selection (see [6] for a full discussion).

The extent to which laboratory-based toxicity tests are, or ever will be, capable of predicting either the likely exposure or the effects of chemical contaminants on ecosystems has also been questioned.[7–11] Little attention has been paid to genotypically or phenotypically determined inter-population differences in susceptibility to contaminant exposure, or to biogeographical differences in chemical toxicity related to climate and the inherent sensitivity of species occupying various regions.[12]

With regard to biomonitoring, the measurement of chemical residues in the tissues of organisms is valuable in assessing bioavailability and in determining the fate of chemicals, but does not provide information about the ecological significance of chemical residues in biota. This raises the question of how relevant biomonitoring studies really are in environmental management.

Ecological studies in which changes in ecosystem structure and function are examined in relation to toxic waste discharges are obviously relevant to the goals of environmental managers. They do not however, afford much practical utility since it is usually difficult to distinguish contaminant-induced ecological change from natural fluctuations, and when such changes are observed, severe deterioration of the ecosystem may already have occurred. Ecological changes can seldom be attributed to a specific cause. Consequently, it is often impossible for environmental managers to take a specific action, such as reducing or banning the discharge of a particular chemical or group of chemicals, because clear evidence of their involvement is not available. The situation may be further confounded by environmental contamination from diffuse sources (such as atmospheric deposition of trace metals and organic compounds) which cannot be controlled in the short term by local measures.

3 VARIABLES OF ECOLOGICAL SIGNIFICANCE

Sibly & Calow[13] provided an excellent summary of the factors which influence the competitive success of organisms, both intra-specifically and inter-specifically. Genes that maximise growth (enabling organisms more rapidly to attain the size at which they can reproduce), and that favour the production of large numbers of viable offspring will be strongly selected. Similarly, genes which enable organisms to survive exposure to environmental stress (natural or anthropogenic) will be favoured. There are trade-offs to be made however, among the advantages conferred by different genes. For example, diverting energy into detoxification and excretory processes may result in slower growth and a fall in reproductive output.[14]

This concept highlights the need for ecotoxicological procedures which include measurements of changes in growth, potential for growth (sometimes measured as Scope for Growth (SFG)[15]), reproductive output, viability of offspring and survivorship resulting from contaminant exposure. It is these variables which influence population dynamics and the competitive success of populations within communities. They are therefore the most relevant variables to measure both in laboratory-based ecotoxicological tests and in field studies.

4 THE USE OF BIOMARKERS TO MONITOR CONTAMINANT EXPOSURE AND EFFECTS

A key role of academic ecotoxicologists is to develop new approaches and procedures that might, in future, become the routine tools of environmental managers. One such approach involves the use of biomarkers (defined below). The biomarker approach has been dogged by spurious claims and by a failure to demonstrate the reliability and cost-effectiveness of biomarker measurements. Studies in Europe and the USA are currently ongoing to address these deficiencies. In the meantime, the potential of the approach should not be ignored.[16]

Ecological change arises due to the differential effects of stressors (including toxicants) on the individuals comprising each population. If a sufficient number of individuals are affected, then changes in population dynamics are likely to have knock-on effects that are manifest as changes in community structure and/or in one or more ecosystem processes.[12]

The identification of specific molecular, biochemical, physiological and behavioural changes in populations of animals and plants following contaminant exposure may provide early warning of impending ecological change.[7,8,16,17,18]

Ecotoxicological biomarkers were defined by Depledge[19] as:

> "Biochemical, cellular, physiological or behavioural variations that can be measured in tissue or body fluid samples or at the level of whole organisms that provide evidence of exposure to and/or effects of, one or more chemical pollutants (and/or radiations)".

The biomarker approach was developed initially to chart the exposure of populations of organisms to contaminants and therefore provided information that was no more ecologically relevant than that derived from biomonitoring studies. However, Depledge,[19,20,21] proposed a new approach in which biomarkers can be used to chart changes in the Darwinian fitness of organisms (Figure 1). It is assumed that a healthy individual exposed to increasing chemical toxicity undergoes a progressive deterioration in health which is eventually fatal. Early departures from health are not apparent as overt disease, but are associated with the initiation of biochemical and physiological compensatory responses. When these compensatory responses are activated, the survival potential of the organism may already have begun to decline because the ability of the organism to mount compensatory responses to new

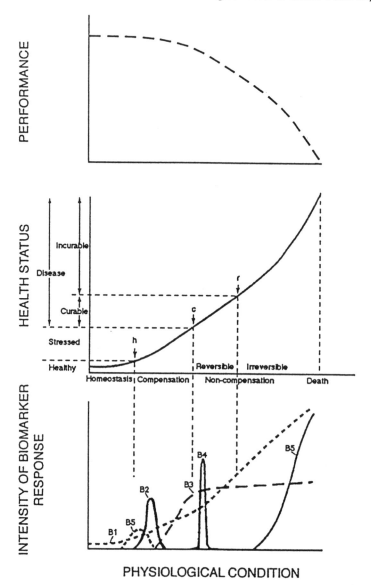

Figure 1   Changes in Performance (Growth, Reproductive Output, Viability of Offspring) in Relation to a Decline in Health Status and Physiological Condition can be Charted with the Aid of Biomarkers,
   h = limit of homeostasis
   c = limit of compensation
   r = limit of recovery.
   B_1–B_5 represent hypothetical biomarker responses that occur with changes in physiological condition.

environmental challenges is compromised. If an organism is exposed to a level of toxicity that cannot be nullified, then pathological processes will result in the development of overt disease and finally, death. Nonetheless, individuals that are no longer able to compensate for the damage caused by the toxicant, may still be able to recover if returned to clean conditions. In other words, an organism can return from a diseased to a healthy state.[19] Biomarkers raise the possibility of determining where an organism is located on this continuum and so potentially provide early warning of reversible, contaminant-induced departures from health (see[20,21] for further details). Furthermore, if compensatory mechanisms utilise energy that would otherwise have been expended on growth and reproduction, then biomarker responses may signal changes in the fitness of organisms that will result in population and community level consequences. The evidence for the existence of such trade-offs is inconclusive. For example, several studies have demonstrated that organisms living in chemically-contaminated conditions have reduced SFGs compared with their counterparts at clean sites (e.g. [22, 23]). This may be interpreted as support for the trade-off hypothesis. However, detailed examination of the data usually (but not always) reveal that reductions in SFG were due to reduced energy intake (reduced feeding) rather than increased maintenance costs associated with compensatory biochemical and physiological responses. Also, it has been estimated that the energetic cost of synthesising proteins associated with detoxification procedures (P450 enzymes, metal-binding proteins, stress proteins, etc.) is relatively trivial and cannot account for the measured reductions in SFG (Garby - personal communication). Further research is required to clarify this issue.

5 DETERMINING ECOSYSTEM STATUS

The term "ecosystem health" is frequently used to indicate that natural ecosystems have a normal set of characteristics which are consistent with a healthy, sustainable status,[24] and that when an ecosystem is contaminated to a sufficient extent, these characteristics change reflecting a decline which is analogous to loss of health in Man associated with the onset of pathological disease processes. It is not the purpose here to offer a full critique of this terminology, although it might be better to restrict the term "health" to descriptions of the state of individual organisms.[25] In the current paper, the term "ecosystem health" is replaced by "ecosystem status", but whatever term is used, the important point to note is that when the characteristics of an ecosystem (structure, processes, species composition etc.) are altered by exposure to anthropogenic chemicals, something has occurred which is of significance to environmental managers. The purist might correctly argue that the new set of ecosystem characteristics are merely different from those in the original ecosystem, and from a scientific standpoint, are no better or worse than the original system. This highlights the partially subjective nature of environmental management. Some ecosystems are more acceptable to human societies than others (although the acceptable forms may be different in different countries). Thus, one of the greatest difficulties faced by environmental managers is to persuade politicians to state precisely which environmental characteristics are acceptable. The outcome of this debate usually involves environmental managers trying to prevent ecosystems changing from their current status as a result of man's activities and, in other cases where environmental quality is obviously poor (overt signs of pollution, low species diversity, an abundance of "weed" species, etc.), trying to restore the environment to a condition similar to that at nearby, relatively clean locations.

From a pragmatic standpoint then, it would be extremely useful if some way could be devised of measuring the changes in ecosystems that are induced by anthropogenic contaminants. Ideally, measurements should signal that components of the ecosystem are exposed to particular classes of chemicals and also indicate when the exposures are associated with changes in biological variables that will ultimately give rise to altered population dynamics, community structure and thereby ecosystem structure and function.

6 THE POTENTIAL ROLE OF BIOMARKERS IN CHARTING CHANGES IN ECOSYSTEM STATUS

The progressive transition of an ecosystem from a relatively stable state to one which is characterised by instability and marked changes in structure and processes (such as nutrient cycling, energy cycling, decomposition rates) is in many ways analogous to the transition of individual organisms from a healthy state to one in which homeostasis fails and pathological processes result in altered energy utilisation. It seems reasonable to assume that the likelihood of an ecosystem changing will increase with increasing chemical contamination (and/or radiation input). However, just as organisms are able to mount compensatory responses to counteract chemical toxicity, the structure and function of ecosystems are preserved in the early stages of contamination due to the phenomenon of "resistance" described by Webster et al.[26] and Harwell et al..[27] A resistant ecosystem may show little change in structure or function if, for example, loss of one or a few populations occurs after contamination. Replacement of lost populations by alternative species that serve the same role maintains ecosystem structure and function.[5] Only when key species are lost, or their well-being is sufficiently impaired, do ecosystem structure and/or function begin to alter. Ecosystem resistance has then been overcome (Figure 2). The replacement of sensitive species in ecosystems by more tolerant ones, without significant changes in ecosystem structure and function might in itself be interpreted as an early warning of an ecologically-significant contaminant impact, especially if loss of the species can be directly attributed to exposure to a particular chemical.

From the above, it is apparent that monitoring changes in sensitive individuals of a given population, or in sensitive populations present within a

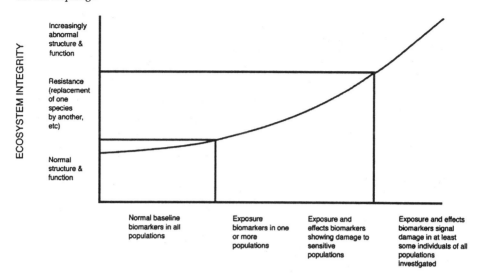

Figure 2 *The Use of Biomarkers to Chart Changes in Ecosystems and Their Component Populations and Communities.*

community, might provide a valuable insight into the status of the whole ecosystem.

7 A PROTOCOL FOR THE USE OF BIOMARKERS IN ECOLOGICAL MONITORING

Depledge and Fossi[12] formulated a protocol for the practical implementation of the biomarker approach alongside the procedures currently used in environmental management. Briefly, it was as follows:

Phase 1   Identification of Ecosystems at Risk

This involves the identification of potential pollutants, pathways and fate together with recognition of critical populations and communities in the ecosystem under study.

A selected suite of general exposure biomarkers would be used in an initial screening of a broad range of species to detect contaminant exposures. (This might include measurement of metallothionein, an indicator of mixed function oxidase activity (BPMO, EROD), cholinesterase activity, a lysosomal integrity assay and a measure of DNA integrity). This would help to establish whether specific contaminants from known point source discharges are initiating biomarker responses or whether unexpected contaminant inputs, perhaps from diffuse sources, are also present.

Phase 2 Identification of Critical Species and Target Populations

Obviously, it is not feasible to perform comprehensive biomarker studies on all the components of an ecosystem. Identification of the most important populations is therefore necessary. An estimate of how much biomarker responses differ among species for a particular level of contaminant exposure may aid in the identification of species at risk. More specific effects of chemicals can be examined using extended suites of exposure and effects biomarkers in a limited range of species occupying different trophic levels and ecological niches.

Phase 3 Extrapolation from Laboratory Studies to Predict the Likely Impact of Chemical Contaminants on Ecosystems

Prediction of the potential of known amounts of specific chemicals to perturb (or further damage) ecosystems can also be aided by the use of biomarkers. Combined laboratory and field biomarker screening tests should be evaluated as a means of establishing a firmer scientific basis for extrapolating from laboratory data to real environments. This might comprise the following:

1) Selection of a range of invertebrate species from diverse phyla that exhibit different feeding strategies and that are present in the ecosystem in question. Sample populations of these organisms should then be exposed to a range of concentrations of the test chemical in the laboratory.
2) Measurement of a suite of biomarkers (biochemical, physiological and behavioural) to assess responses to, and toxicity of the test chemical should then be performed. Biochemical biomarkers should reveal the type of detoxification mechanisms induced by the chemical whilst physiological and behavioural biomarkers will signal exposures resulting in adverse effects at the level of the whole organism (such as altered scope for growth, loss of endogenous behavioural rhythmicity, etc.). They will also permit time relationships between exposure and biomarker responses to be established.
3) Residue analysis of the test organisms should be carried out to relate biochemical biomarker responses in specific tissues to tissue concentrations of the test chemical or its derivatives.
4) If the test chemical has been released into other similar ecosystems and biomarker responses have been measured *in situ*, then results obtained in the laboratory test can be compared with the database compiled from field tests.
5) Once the test chemical has been evaluated and safe concentrations determined, the biomarker approach offers the possibility of genuine validation of the test procedure. Verification that the chemical in question, at a given concentration, induces the same biomarker response in the field as it does in the laboratory would be particularly helpful. If not, then there is a good basis for establishing whether organisms *in situ* are more or less sensitive to the chemical in the field than they were in the laboratory, and by how much.

Establishment of a database of biomarker responses from diverse studies on a range of organisms subjected to various degrees of contaminant exposure, both in the laboratory and *in situ*, will allow the extent of biomarker responses to be related to general levels of ecological change. If contaminant concentrations *in situ* are then held below concentrations that do not produce adverse biomarker responses in a significant proportion of the test population, protection of the ecosystem should be ensured.

8 SUMMARY

The biomarker approach permits acquisition of information that cannot be obtained from the measurement of chemical residues in environmental and biological media. Biomarkers enable integration of pharmacokinetic and toxicological interactions in an organism resulting from exposure to complex mixtures of chemicals; they indicate the cumulative effects of toxicant interactions in molecular or cellular targets and they integrate different episodes of exposure in time and space. Different biomarkers are capable of indicating rapid responses to contaminant exposure, or of providing early warning of long-term effects on populations, communities and whole ecosystems. The significance of chemical residues in tissues can only be assessed if detailed toxicological studies have been carried out on the target organism, relating residue levels to biomarker responses and changes in development, growth, reproductive output, viability of offspring and survivorship.

In the design of new ecotoxicological test procedures it is recommended that:

(i) More attention should be given to measuring changes in ecologically-**SIGNIFICANT** variables in relation to contaminant exposure.
(ii) Results of laboratory test procedures should be validated by *in situ* measurements to ensure their **RELEVANCE**.
(iii) The feasibility of **EXTRAPOLATING** from measurements of biomarkers of adverse effects in individuals to predict contaminant-induced changes in populations and communities, should be explored.
(iv) The feasibility of charting changes in ecosystems induced by specific toxic waste discharges using biomarker measurements in sentinel populations should also be considered.

ACKNOWLEDGEMENTS

The author is grateful to the European Commission and the Danish Government (Strategic Environmental Research Programme) for funding the research which has allowed the concepts and approaches outlined in this paper to be developed.

References

1. T. H. Maugh, *Science,* 1978, **199**, 162.
2. N. B. Chapman and J. Shorter, Advances in Linear Free Energy Relations, Plenum Press, London, 1972.
3. A. M. V. M. Soares and P. Calow (Eds), Progress in Standardisation of Aquatic Toxicity Tests, Lewis Publishers, Boca Raton, 1993.
4. D. J. H. Phillips and P. S. Rainbow, Biomonitoring of aquatic trace contaminants, Chapman and Hall, London, 1993.
5. J. R. Kelly and M. A. Harwell, in Ecotoxicology: Problems and Approaches, edited by S. A. Levin, M. A. Harwell, J. R. Kelly & K. D. Kendall, 1989, Springer-Verlag, New York, 9-35.
6. F. Moriarty, "Ecotoxicology: the study of pollutants in ecosystems", Academic Press, Cambridge, 1983.
7. M. H. Depledge, Proceedings of the 12th Baltic Marine Biologists Symposium, edited by E. Bjørnstad, L. Hagerman, & K. Jensen, Olsen & Olsen, Fredensborg, pp. 47-52, 1992.
8. J. Cairns, Jr. and P. V. McCormick, *The Environmental Professional,* 1992, **14**, 186.
9. J. Cairns, Jr. *Hydrobiologia,* 1983, **100**, 47.
10. R. A. Ryder and C. J. Edwards, Report to the Great Lakes Science Advisory Board, 1985, Windsor, Canada.
11. K. D. Kimball and S. A. Levins, *Bioscience,* 1985, **35**, 165.
12. M. H. Depledge and M. C. Fossi, *Ecotoxicology,* 1994, **3**, 161.
13. R. M. Sibly and P. Calow, Physiological Ecology of Animals, Blackwell Scientific Publication, Oxford, 1986.
14. A. A. Hoffman and P. A. Parsons, Evolutionary genetics and environmental stress, Oxford Science Publishers, Oxford, 1991.
15. S. Brody, "Bioenergetics and growth", Reinhold Publishing Corporation, New York, 1945.
16. J. F. McCarthy and L. R. Shugart, (Eds), Biomarkers of Environmental Contamination, Lewis Publishers, Boca Raton, 1990.
17. D. B. Peakall and L. R. Shugart, Biomarkers: research and application in the assessment of environmental health, Springer-Verlag, Berlin & Heidelberg, 1992.
18. M. C. Fossi, and C. Leonzio, Nondestructive Biomarkers in Vertebrates, 1994, Lewis Publishers. Boca Raton, 1994.
19. M. H. Depledge, in "Nondestructive Biomarkers in Vertebrates", 1994, edited by M. C. Fossi and C. Leonzio, Lewis Publishers. Boca Raton, 1994.
20. M. H. Depledge, *Ambio,* 1989, **18**, 301.
21. M. H. Depledge, J. J. Amaral-Medes, B. Daniel, R. Halbrook, P. Kloepper-Sams, M. N. Moore and D. B. Peakall, in "Biomarkers: Research and Application in the Assessment of Environmental Health", edited by D.B. Peakall & L.R. Shugart, NATO ASI Series H: Cell Biology, **68**, Springer-Verlag, Berlin, Heidelberg, 1992.
22. L. Maltby, *Environ. Toxicol. Chem.,* 1992, **11**, 79.
23. J. Widdows, K. A. Burns, N. R. Menon, D. S. Page, and S. J. Soria,. *Exp. Marine Biol. Ecol.* 1990, **138**, 99.
24. R. T. Di Giulio and E. Monosson, Interconnections between human and ecosystem health, Chapman and Hall, London, 1996.

25 G. W. Suter, III, *Environ. Toxicol. Chem.*, 1993, **12**, 1533.
26 J. R. Webster, J. B. Waide and B. C. Patten, in "Mineral cycling in southeastern ecosystems", edited by F.G. Howell, J.B. Gentry and M.H. Smith, ERDA, Springfield, Virginia, 1975.
27 M. A. Harwell, W. P. Cropper and H. L. Ragsdale, *Ecology*, 1978. **58**, 660.

In situ Assays for Monitoring the Toxic Impacts of Waste in Rivers

M. Crane[1], I. Johnson[2] and L. Maltby[3]

[1]Division of Biology, Royal Holloway University of London, Egham, Surrey TW20 0EX, UK
[2]WRc plc, Henley Road, Medmenham, Marlow, Buckinghamshire SL7 2HD, UK.
[3]Department of Animal and Plant Sciences, The University of Sheffield, Sheffield S10 2TN, UK

ABSTRACT

Environmental regulators world-wide have increased their use of laboratory bioassays for assessing the toxic effects of chemicals. This can be a cost-effective approach with many advantages over the traditional reliance upon chemical-specific standards. However, the extrapolation of hazards found in laboratory bioassays to risk of damage in the natural environment can be difficult. One reason for this is that exposure conditions in the laboratory may not adequately reflect true exposure conditions in the field. The use of site-specific, *in situ* assays can help to overcome this problem.

Field-based *in situ* tests have several advantages over laboratory-based techniques. Toxic episodes may be missed if water is collected as a 'spot' sample and then transported to the laboratory for testing; and toxicants that volatilise, degrade or partition rapidly onto solids may be overlooked if there is a delay between sample collection and testing. *In situ* assay organisms are exposed to toxicants under more natural conditions, thus integrating site-specific effects over time. *In situ* assays can be used within a survey-based or experimentally-based strategy, and can provide a link between laboratory predictions and field effects. Various measurement parameters are available, spanning the range of biological organisation from biochemical to community indices.

This paper describes several studies in which amphipods have been used *in situ* to assay the toxicity of wastes in rivers. The effects of sewage, industrial and agricultural wastes on amphipod mortality, physiology and biochemistry are discussed. A framework is presented for the complementary use of such *in situ* assays within current and proposed river quality assessment procedures.

1 INTRODUCTION

In recent years, environmental regulators world-wide have increased their use of laboratory bioassays for assessing the toxic effects of chemicals. This can be a cost-effective approach with many advantages over the traditional reliance upon chemical-specific standards, especially when waste effluents are complex.[1] Analytical chemistry, when used alone, has the following disadvantages:[2]

(i) many effluents contain organic chemicals that are not easy to identify or measure;
(ii) toxicological data are missing for many synthetic chemicals, so even if their levels can be measured their likely biological effect remains unknown;
(iii) interactions between the individual components of a complex effluent that affect its toxicity cannot be measured chemically;
(iv) measurement of all the constituents of a complex effluent, where this is possible, is expensive.

The use of bioassays with fish, invertebrates and plants gives a direct assessment of the toxicity of an effluent. Direct Toxicity Assessment (DTA) can then be used to formulate a Toxicity Based Consent (TBC) that obliges the waste discharger to produce effluent that falls below a quantifiable level of toxicity.[2]

However, the extrapolation of hazards found in laboratory bioassays to risk of damage in the natural environment can be difficult. One reason for this is that exposure conditions in the laboratory may not adequately reflect true exposure conditions in the field. The use of site-specific, *in situ* assays in combination with laboratory assays and field monitoring can help to overcome this problem.[3]

This paper first describes some of the advantages of *in situ* bioassays and then provides several examples of their use in rivers. The amphipod *Gammarus pulex* L. is used as a model organism because a considerable amount of work has been done with this animal at several levels of biological organisation. The paper concludes with some suggestions on how *in situ* assays can usefully be integrated into Direct Toxicity Assessment programmes.

2 ADVANTAGES OF *IN SITU* BIOASSAYS

In situ assays are site-specific monitoring exercises in which the investigator either samples naturally-occurring populations, or transplants organisms from elsewhere. Structural or functional measurements can be made on endpoints such as species richness and abundance, lethal and sub-lethal toxic effects, and chemical uptake.[4]

Field-based *in situ* tests have several advantages over laboratory-based bioassay techniques. Toxic episodes may be missed if water is collected as a 'spot' sample and then transported to the laboratory for testing; toxicants that volatilise, degrade or partition rapidly onto solids may be overlooked if there is a delay between sample collection and testing; and *in situ* assay organisms are exposed to toxicants under more natural conditions, thus integrating site-specific exposure over time.[3] *In situ* assays can be used within a survey-based strategy to monitor the spatial and temporal distribution of toxicity,[3] or within an experimentally-based strategy to test specific hypotheses.[9] Within both strategies they can provide a link between laboratory predictions and field effects.

Many different *in situ* bioassays can be used to assess river quality. Macroinvertebrate species richness and abundance can be simply monitored using pond nets, Surber samplers or colonisation units.[3] Fish species richness and abundance can also be monitored by using electrofishing or hydro-acoustic techniques. These are the traditional tools of the river biologist responsible for monitoring the quality of receiving waters. However, the emphasis in this paper is not on surveys of natural populations, invaluable as they are, but on the use of caged organisms placed experimentally in receiving waters. The main advantages of this approach are that covarying factors are more easily controlled and the possibility of pollution-induced tolerance is avoided. The amphipod *Gammarus pulex* L. has been deployed extensively in this way and can be used to illustrate the strengths of the approach.

3 *IN SITU* ASSAYS WITH *GAMMARUS PULEX*

Gammarus can be deployed in the field in small plastic cages capped with 1 mm mesh to allow the flow of water through the bore of the cage.[5] These are then placed *in situ* with or without food. Because of their cannibalistic tendencies, *Gammarus* should normally be kept individually, unless the study is of only a few hours duration.

3.1 Mortality

The simplest approach to investigating toxicity with these animals is to look at lethal effects, since death is an easy endpoint to measure. *Gammarus* mortality was investigated in a small river catchment in Kent subject to sewage and agricultural wastes and natural contamination from iron-rich groundwater.[3] Thirty to forty animals were placed at each of six or seven stations on two separate occasions: once during Summer and again in the following Spring. Deployments lasted for 28 days. Mortality was assessed at the end of this

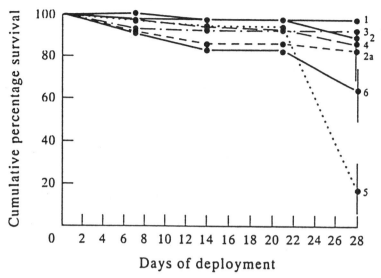

Figure 1 *Survival of caged* Gammarus *at seven stations in an agricultural catchment (Spring deployment). Error bars are 95% confidence intervals.*

period on both occasions, and also at intervals during this period in the second study.

During the first deployment, in the Summer, there was complete mortality of gammarids deployed below the point of entry of sewage effluent from a Young Offenders' Institution (stations 3-6). There was a different pattern of mortality during the subsequent Spring deployment. Most animals survived for the first three weeks of the study, but within the last week most died at station 5 near agricultural land recently sprayed with pesticides (Figure 1). There was also some mortality at station 6, near the confluence with the main river in the catchment.

There was no evidence for poor water quality from chemical analysis of water samples in either of these studies, and the highly variable macroinvertebrate data collected from the stream were also difficult to interpret. The *in situ* bioassays with *Gammarus* clearly showed points in the stream where acute lethal toxicity occurred and gave some indication that sewage waste was a likely cause in the first study, and that agricultural wastes were a likely cause in the second study.

Total mortality at a fixed time interval is not the only way in which survival data can be used. *Gammarus* were deployed in the River Calder, Yorkshire, above and below Huddersfield Sewage Treatment Works for ten days. By the end of this period all of the fifty animals placed upstream and all fifty placed downstream were dead (Figure 2). This is not surprising since the Calder is contaminated by sewage and industrial wastes both above and below the Sewage Treatment Works outfall. However, gammarids placed below the outfall died at a more rapid rate than those above the outfall, indicating that water quality below the outfall was worse.

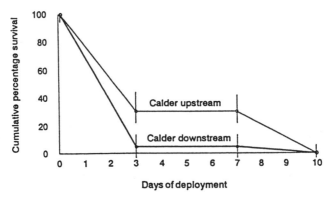

Figure 2 Survival of caged Gammarus in the River Calder above and below Huddersfield Sewage Treatment Works outfall. Error bars are 95% CI.

3.2 Feeding Rate

Mortality is rather a coarse measure of toxicity, so several sub-lethal measurements have been developed for use *in situ*. Feeding rate is the most widely used of these sub-lethal endpoints because it can be measured simply, with reasonable accuracy and precision.[5,6,7] The caged gammarids are supplied with a known weight of food material, usually alder (*Alnus glutinosa* L.) or horse chestnut (*Aesculus hippocastanum* L.) leaves. They are then deployed in the field with this food for a few days or weeks. After deployment, the animals are killed, dried and weighed and the remaining food material is also weighed. The feeding rate of the animals is then calculated as the weight of food consumed, normalised for the weight of the animal.

This test was deployed during Summer and Spring above and below a yarn dying works suspected of the episodic release of pyrethroid insecticides and dyes into the River Tame in Lancashire. There was no effect on *Gammarus* feeding rate during the first deployment. However, gammarids below the works fed less on the second deployment (ANCOVA $F = 20.05$, $p = 0.021$; Figure 3), and this reduction in feeding rate was associated with a reduction in macroinvertebrate biotic score. This was an indication that a pollution episode had occurred during the period of deployment.

In a similar way to the mortality data described earlier, changes in amphipod feeding rate may also provide a sensitive indication of reduced water quality when macroinvertebrate monitoring provides only equivocal results. In a longitudinal study down the River Aire in Yorkshire, it was found that despite a reasonably high macroinvertebrate biotic score for a station below Gargrave Sewage Treatment Works, *Gammarus* feeding rate was significantly reduced (Figure 4). This type of information from caged organism studies complements the analytical chemistry and macroinvertebrate monitoring studies that are currently performed by environmental regulators.

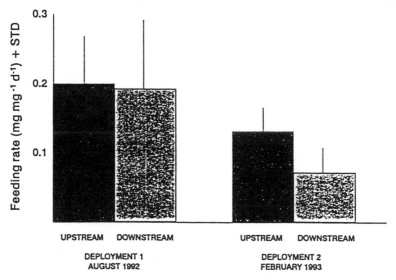

Figure 3 *Feeding rates of caged* Gammarus *above and below a dye works during two deployments. Error bars are one standard deviation.*

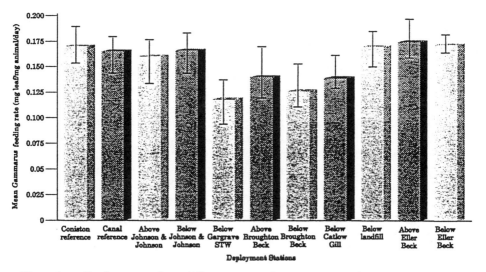

Figure 4 *Feeding rates of caged* Gammarus *at eleven stations in the River Aire. Error bars are 95% CI.*

3.3 Biomarkers

Toxic effects on survival and physiology may take some time to manifest themselves. Effects at the molecular and biochemical level are likely to occur more rapidly, or at least to be measurable at an earlier stage. This is one reason why biomarkers of exposure or effect are currently a popular subject

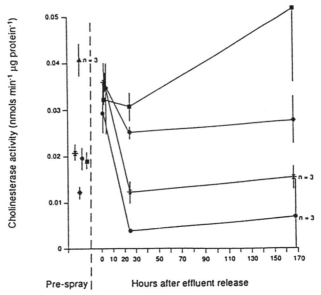

Figure 5 *Cholinesterase activity in caged* Gammarus *deployed at four field stations before and after application of Malathion 60 to watercress. Error bars are standard errors; n=4 except where marked otherwise. Animals placed immediately below watercress beds = * and •; animals placed further downstream = ♦ and ■.*

for ecotoxicological research.[8] Experimental *in situ* assays are a good vehicle for these newer techniques because their use avoids problems of possible resistance or tolerance in natural populations.

The inhibition of acetylcholinesterase (AChE) in *Gammarus* can be an effective way of monitoring *in situ* exposure to organophosphorus pesticides.[9] Caged *Gammarus* placed immediately below watercress beds sprayed with malathion, an organophosphorus insecticide, showed a significant increase in AChE inhibition (Figure 5). Those deployed further down the waste stream showed no significant inhibition. This, with information from other studies, indicated that the waste control measures adopted by the watercress growers were adequate.

4 THE ROLE OF *IN SITU* ASSAYS IN DIRECT TOXICITY ASSESSMENT

The examples given above illustrate the availability of experimental *in situ* techniques with organisms like *Gammarus* at different levels of biological organisation. But are such *in situ* techniques of practical relevance in the context of waste regulation, Direct Toxicity Assessment and Toxicity Based Consents?

Proposals for implementing DTA in this country recommend a four stage procedure. A potential candidate for DTA enters Stage 1, the Selection and Prioritisation Stage. A desk-based appraisal is carried out to define the receiving water mixing zone and to determine the effluent concentration at the edge of the zone. This information is used in conjunction with information from simple, rapid screening tests, such as Microtox™, to determine whether a TBC would be appropriate for the effluent. If a TBC is deemed appropriate, then the effluent passes to Stage 2, an in-depth assessment of toxicity. In this stage, both simple bioassays and higher organism bioassays with fish, invertebrates and algae are used to measure toxicity, and the effluent concentration that has no adverse effect at a designated point in the receiving water is estimated. If the effluent is likely to be toxic at this point then its toxicity must be reduced before it is consented for discharge. Stage 3 is the derivation of the TBC. This will be based upon either a simple screening test, if this can be calibrated with a higher organism test or, if such a calibration is not possible, upon the most sensitive higher organism test. Stage 4 comprises monitoring to ensure compliance with the TBC and regular reviews to establish its effectiveness.

Experimental *in situ* bioassays could play a useful role in Stages 2 and 4 of DTA. In Stage 2, the spatial mapping of site-specific toxicity above and below a discharge would be a useful way of testing predictions based on laboratory bioassays and desk-based estimates of mixing zones. This sort of mapping exercise would also show the *additional* toxicity contributed by a particular effluent to a receiving water that is already contaminated by upstream effluents, and provide important information for a cost-benefit analysis.

In Stage 4, the use of experimental *in situ* assays for the temporal mapping of site-specific toxicity above and below a discharge would be a useful way of monitoring any improvements in receiving water quality after a TBC has been set. Recovery in biochemical or physiological activity is likely to occur more rapidly than recovery in fish or macroinvertebrate species richness or abundance. Experimental *in situ* assays will give an early, and potentially more precise, indication of whether the TBC is an effective solution to effluent toxicity.

5 CONCLUSIONS

Experimental *in situ* bioassays with organisms like *Gammarus* have been used extensively to monitor the site-specific effects of many different types of waste. Their advantages over more traditional biomonitoring methods are that they can produce more precise measurements, and tolerance or resistance to contaminants can be excluded. It is also easier to control other covarying factors, such as flow rate and substrate characteristics. Although site-specific field studies such as this are necessarily pseudoreplicated, as is all river monitoring, experience has shown that they can provide a strong indication of the possible sources and effects of effluent contamination.

Gammarus is not the only invertebrate suitable for field deployment. Recent work with the freshwater midge *Chironomus* has shown that this sediment-dweller can also be deployed successfully *in situ,* and that a similar range of endpoints to those used with *Gammarus* can be measured in these insect larvae. Even animals as small as water fleas can be placed in the field. Structured population models that use information on survival, growth and development can then be applied to data from caged *Gammarus*, *Chironomus* and *Daphnia* to provide predictions of population-level effects.

In the context of DTA, experimental *in situ* bioassays can help both regulators and waste dischargers by reducing the uncertainty associated with extrapolations of toxicity from the laboratory to the field. More accurate spatial and temporal mapping of site-specific toxicity is possible when several *in situ* and laboratory bioassays with similar organisms are combined with chemical analyses and field monitoring of species richness and abundance in an integrated approach. This allows the regulator to set a realistic yet environmentally protective TBC, and both the regulator and the waste discharger to monitor the quantitative benefits of that TBC.

References

1. National Rivers Authority, 'Discharge Consent and Compliance Policy: a Blueprint for the Future.' Water Quality Series No. 1, NRA, London, 1990.
2. I. Johnson, R. Butler and N. Adams, 'The Control of Effluent Discharges by a Direct Toxicity Assessment (DTA) Approach.' Report to the National Rivers Authority, 1993.
3. M. Crane, P. Delaney, C. Mainstone and S. Clarke, *Wat. Res.*, 1995, **29**, 2441.
4. S. P. Hopkin, In, 'Handbook of Ecotoxicology', edited by P. Calow, Blackwell, Oxford, pp 397-427, 1993.
5. M. Crane and L. Maltby, *Environ. Toxicol. Chem.*, 1991, **10**, 1331.
6. J. Seager, I. Milne and M. Crane, In, 'Ecological Indicators Vol. 1', edited by D. H. McKenzie, D. E. Hyatt and V. J. McDonald, Elsevier, London, pp 243-258, 1992.
7. J. Seager, I. Milne, G. Rutt and M. Crane, In, 'River Water Quality Ecological Assessment and Control', edited by P.J. Newman, M.A. Piavaux and R.A. Sweeting, Commission of the European Communities, Brussels, pp 399-416, 1992.
8. L. R. Shugart, J. F. McCarthy and R. S. Halbrook. *Risk Analysis*, 1992, **12**, 353.
9. M. Crane, P. Delaney, S. Watson, P. Parker and C. Walker, *Environ. Toxicol. Chem.*, 1995, **14**, 1181.

Sub-lethal Biological Effects Monitoring Using the Common Mussel (*Mytilus edulis*): Comparison of Laboratory and *In situ* Effects of an Industrial Effluent Discharge

B. D. Roddie,[1] C. J. Redshaw[2] and S. Nixon[3]

[1]Environment and Resource Technology Ltd, Waterside House, 46 The Shore, Edinburgh EH6 6QU, UK
[2]SEPA West Region HQ, Rivers House, Murray Road, East Kilbride G75 0LA, UK
[3]WRc plc, Henley Road, Medmenham, Bucks. SL7 2HD, UK

ABSTRACT

The study reported here formed part of an extensive programme of 17 field studies designed to evaluate *in situ* sub-lethal methods and was carried out between 1985 and 1989. Mussels were deployed in the vicinity of an industrial discharge for a period of four weeks, following which the cumulative effects of exposure on feeding and respiration rate were measured in the laboratory. Concurrently, composite samples of the effluent were collected at weekly intervals, and mussels were exposed in the laboratory to serial dilutions of this effluent. The feeding rates of laboratory-exposed mussels were measured at weekly intervals following temporary transfer to clean seawater.

Comparison of the responses of field- and laboratory-exposed mussels indicated good agreement between the effects of actual and estimated dilutions of effluent within the mixing zone. Laboratory studies also indicated a marked change with time in effluent composition which was clearly reflected in week-to-week variation in mussel feeding rate.

This study indicated, by means of integrated field and laboratory studies, that sub-lethal stress measurement in mussels is a practical tool for assessing effluent and receiving water quality.

1 INTRODUCTION

Toxicity tests are being used increasingly to assess the potential biological impact of industrial discharges and to derive consent conditions for their control. Data derived from substance specific toxicity tests are often also the primary basis on which Environmental Quality Standards (EQSs) are established. Many tests used in this context are acute (i.e. short-term) tests, and have lethal endpoints. The use of such tests is essential to support the process of whole-effluent toxicity assessment, and to permit the establishment of a defensible compliance monitoring scheme. However, acute, lethal toxicity tests provide a less adequate basis for control as regulators demand increasingly stringent standards; their limitations become even more apparent when attempts are made to extrapolate from effluent test data to predictions of receiving water quality.

In setting consents and deriving EQSs, it is desirable to incorporate at some point an element of validation. These standards are established with the intention of conferring a degree of protection on the receiving environment, but there is, often, little or no provision for checking that the expected effects on water quality are in fact realised. Since extrapolation from toxicity test data to 'no effect' predictions invariably requires the use of approximations and safety factors, it can be difficult to demonstrate to the satisfaction of all parties that the limits which are established are neither grossly over- or underprotective.

Sub-lethal biological effects test methods, and especially those which can be deployed in receiving waters, offer the potential to address the above limitations. The use of sub-lethal endpoints requires fewer assumptions regarding concentration (exposure)-response relationships, and (in principle at least) permits a more realistic estimation of no-effect concentrations. *In situ* exposure, similarly, requires fewer assumptions about the fate and behaviour of contaminants, and can provide a means of integrating effects over longer periods than are conventionally examined in the laboratory. Approaches to *in situ* evaluation include the transport of ambient water to the laboratory for lethal and sub-lethal testing[1] as well as the deployment of both marine and freshwater organisms at locations impacted by anthropogenic discharges.[2] *In situ* exposure, however, may be inherently more likely to enable the detection of effects due to volatile or unstable chemicals, which may achieve equilibrium concentrations in receiving water, but which may be readily lost from samples abstracted from the receiving environment.

The marine mussel, *Mytilus edulis*, has been used widely in studies of contaminant accumulation ('mussel watch' programmes), and has also been shown to have considerable potential for the assessment of sub-lethal biological effects.[3] Mussels are easily deployed in field situations, and are amenable to a range of physiological rate measurements which can respond sensitively to a wide range of contaminant types.[4] Feeding and respiration rates have been widely-used in this respect, and, when expressed in common energy terms, constitute the primary components of Scope for Growth (SFG). SFG represents the net difference between energy assimilation and energy expenditure, and thus provides an indication of the resources available for growth and reproduction.

The study described here was one of a series of field trials designed to evaluate SFG and its components as a means of assessing *in situ* receiving water quality.

2 MATERIALS AND METHOD

2.1 General Design

The objective of the study was to investigate the *in situ* sub-lethal impact of an industrial discharge, in parallel with more direct assessment of the sub-lethal toxicity of the effluent in the laboratory.

2.2 Field Study Design

The batch of mussels used in this study were obtained from a natural population in the Wash. These animals were acclimated to laboratory conditions over a two week period prior to deployment in the field. During this period, preliminary physical, physiological and chemical measurements were made.

Mussels were deployed on 7 July 1988 in cages (approximately 100 per cage) attached in pairs to eight buoy systems positioned at outfall and reference locations in Irvine Bay, Ayrshire, Scotland. Six of these buoys were positioned in pairs to the north and south of the industrial outfall and an adjacent sewage outfall, all within 100 m of the end of the diffuser sections; the positions were selected to place the mussels well within the permitted 'mixing zones', which had an estimated radius of 300-400 m. The remaining two buoys were positioned approximately 15 km to the south, to serve as remote reference points with similar hydrography.

One objective was to attempt to distinguish the effects of the industrial outfall from the effects of the adjacent sewage outfall. The mussels were retrieved on the 16, 18 and 22 August 1988. To allow for possible day-to-day variation in laboratory measurement conditions, a sample of mussels from the reference site (buoys seven and eight) was retrieved on each day and the reference values thus derived were used to standardise between-site comparisons.

Groups of mussels retrieved from the field were transported to Clyde River Purification Board (CRPB) (now SEPA) laboratories, and were processed within 24 hours of collection. Mussels were placed in cool boxes directly following retrieval; care was taken to maintain them in a cool, humid environment at all times during collection and transport. The interval between removal from site and arrival at the laboratory was less than six hours.

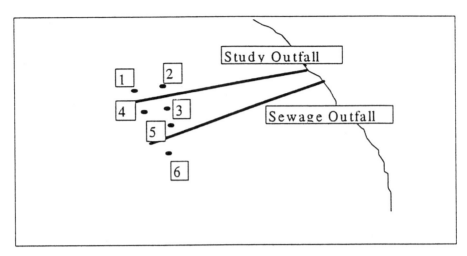

Figure 1 *Approximate Relative Locations of Mussel Buoys*

Measurements were conducted on groups of 11 mussels from each field site.

2.3 Laboratory Study Design

The effluent discharger collected seven composite 24 hours samples per week, and stored these at approximately 6 °C in the dark. At weekly intervals, these samples were combined and shipped in cool boxes to the laboratory. Mussels were exposed in groups of eight in 10 litre tanks to concentrations of 0, 0.08, 0.15, 0.25, 0.45, 0.8 and 8% effluent in filtered seawater. Exposure was semi-static, with medium replacement at 24 hours intervals. Gentle aeration was supplied continuously to all exposure tanks. Mussels were removed from the exposure tanks, and their feeding rate was measured as described below, at weekly intervals.

The effluent discharger performed chemical analysis on each weekly effluent composite sample, and reported the results in terms of trace metal concentration and the concentration of specific consented organic solvents.

2.4 Physiological Measurements

Although one objective of the study was to evaluate the use of SFG in field monitoring, for practical reasons only two physiological properties were measured; feeding rate and respiration (oxygen consumption) rate. Other factors which, in principle, influence energy assimilation and expenditure rates (such as food absorption efficiency and nitrogen excretion rate) were considered to be either negligible or to be subject to a substantial degree of error; in calculating SFG, standard values were assigned to these factors. Food energy content was also assigned a standard value, because measurements were made under controlled laboratory conditions; no quantitative data were available for actual ration quantity or quality over the period of field deployment.

Following the recommendations of Bayne et al,[4] all measurements were made on clearly-marked individuals, to permit individual energy budgets to be constructed. In order to achieve this, all mussels selected for measurement were marked immediately upon arrival in the laboratory with an indelible, non-toxic white paint. This approach also permitted the effects on mussels in laboratory effluent tests to be tracked on an individual basis over time.

2.4.1 Feeding Rate. Feeding rate was measured in a flow-through system on an individual basis. For pre-deployment measurements, filtered seawater from a recirculating treatment system was supplied at approximately 15 °C. For post-deployment measurements, seawater collected from a clean offshore site was supplied from a 1000 litre holding tank located in a controlled-temperature room (at field ambient temperature +/- 1 °C). In each instance, water was delivered to individual 500 ml vessels at a rate of approximately 150 ml min^{-1}.

The seawater was supplied via a mixing chamber, into which a concentrated culture of the marine diatom *Phaeodactylum tricornutum* was injected by peristaltic pump to give an outflow concentration of approximately 3000 cells ml^{-1}. Mussels were placed individually in the vessels, positioned with the inflow directed towards their inhalant siphons, and allowed to acclimate to experimental conditions for two hours. This acclimation period also served to allow the 'repayment' of any oxygen debt accumulated during transportation from the field to the laboratory. At least one vessel on each measurement occasion did not receive a mussel; this vessel was used as a control, to estimate the concentration of algae in the inflowing water.

On three occasions at hourly intervals after acclimation, outflow samples were collected simultaneously from all vessels over a period of approximately one minute. *Phaeodactylum* cell concentrations were measured within 30 minutes of collection on three well-mixed aliquots of each sample, using a Coulter Counter model TAII, equipped with a 140 (m orifice tube.

The average cell concentration for each sample was calculated, and the feeding rate (as clearance rate) calculated for each sampling time from the expression:

$$CR(l\,h^{-1}) = \frac{[Inflow] - [Outflow]}{[Inflow]} \times Flowrate(l\,h^{-1})$$

where [Inflow] and [Outflow] represent *Phaeodactylum* concentration in cells ml^{-1}. The final clearance rate value for each individual was expressed as the average of the three values obtained over the measurement period of two hours.

During the entire measurement period, but more particularly during the periods of overflow sample collection, care was taken to avoid sudden changes in light, shading, noise and vibration, since these had been observed in previous studies to elicit sudden closure responses in mussels.

2.4.2 Respiration Rate. Respiration rates were measured as oxygen consumption in individual respirometry chambers constructed from 250 ml Sartorius polycarbonate filter receiving vessels. Dissolved oxygen concentrations were measured using Radiometer microcathode electrodes connected via Strathkelvin Instruments amplifiers to four-channel chart recorders. The respirometry chambers were placed in a temperature-controlled water bath supported on a set of magnetic stirrer units. Each chamber contained a magnetic follower retained within an ABS ring attached to the base; a mussel was placed individually within the chamber to one side of the retaining ring, which served to isolate it from mechanical disturbance.

Each vessel was fitted with a domed lid, with a screw fitting sealed by a silicone rubber 'O' ring. The domed lid facilitated the exclusion of air bubbles while filling the vessel. When filling was complete, an electrode was inserted into the central aperture; the electrode body was wrapped with Parafilm to ensure an airtight fit.

The decline in oxygen concentration within the respirometer was recorded for 30-45 minutes for each mussel, and the rate of consumption (ml O_2 h^{-1}) was calculated by fitting a straight line by eye to the 'linear' section of each chart recorder trace.

2.4.3 Physical Measurements. In the field study, shell length, shell volume, and wet and dry tissue weight were recorded before and after deployment.

3 RESULTS

3.1 Field Study

Mussels were successfully deployed and retrieved from all sites except site three (to the south of the industrial outfall).

All physiological rates were allometrically corrected to a standard 1.0 g dry flesh weight, to remove the effect of variation in individual size. In calculating SFG, a standard food energy absorption efficiency of 0.6 was assigned, and a standard ration energy value of 16 J per litre of water filtered was used. Energy assimilation rates are thus directly proportional to feeding (clearance) rates. Oxygen consumption rates were converted to energy expenditure rates assuming mixed metabolism.

Survival of mussels was good, with a maximum observed mortality of 4% in one of the reference site cages. Mussel cages from the reference site were heavily colonised with hydroid growth, while cages retrieved from the vicinity of the outfall were markedly less well-colonised.

Table 1 *Average Energy Assimilation and Expenditure Rates ($J\ g^{-1}\ h^{-1}$), (95 % CI)*

Site	Assimilation	Expenditure	SFG
Pre-deployment	24.5 (3.0)	7.6 (0.6)	16.7 (3.2)
1 (a)	14.8 (3.1)	11.13 (1.8)	4.1 (2.9)
2 (a)	14.8 (3.6)	12.06 (1.8)	2.6 (3.2)
4 (b)	17.5 (2.8)	9.16 (1.1)	8.2 (2.4)
5 (c)	24.0 (1.6)	9.26 (1.4)	14.7 (1.5)
6 (b)	19.6 (3.5)	9.46 (0.9)	10.0 (3.7)
7 (a)	18.3 (2.7)	8.37 (1.1)	9.9 (2.6)
7 (b)	18.1 (3.8)	6.83 (0.4)	11.2 (3.8)
7 (c)	26.2 (2.4)	5.05 (0.4)	21.1 (2.3)
Tukey's HSD	7.07	2.61	4.94

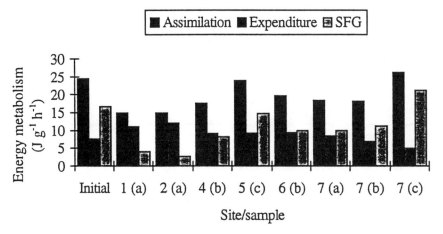

Figure 2 *Energy Metabolism Rates in Field-Deployed Mussels*

Energy assimilation and expenditure rates ($J\ g^{-1}\ h^{-1}$) are summarised in Table 1 and Figure 2. Mussels were retrieved from one of the reference sites (site seven) on each retrieval occasion, and the exposure sites associated with each reference site sample are designated a), b), or c) accordingly.

These data were analysed by one-way ANOVA, and differences between means assessed by computing Tukey's Honestly Significant Difference (HSD). Sites one, two, and four exhibited the lowest feeding rates and SFG, and the highest respiration rates. There was clear evidence of significant day-to-day variation from a comparison of the three reference site values, and to remove the effect of this the average SFG values for each site were adjusted relative to a notional reference site value of 10 $J\ g^{-1}\ h^{-1}$ (Table 2).

These adjusted values distinguish sites one and two clearly from the remaining sites, but suggest that there is relatively little real difference between sites four, five and six.

Both feeding rate and respiration rate exhibited marked variation between

Table 2 *Relative SFG Values for Exposure Sites, Adjusted for Variation in Control Site Values*

Site	Adjusted SFG	Location
1	4.1	N of outfall
2	2.6	N of outfall
4	7.3	S of outfall
5	6.9	S of sewer
6	8.9	S of outfall

Table 3 *Average (95% CI) Body Condition Index and Dry Flesh Weight*

Site	Body condition index	Dry tissue weight (g)
Pre-deployment	79.2 (4.03)	0.61 (0.09)
1 (a)	159.9 (14.1)	1.32 (0.11)
2 (a)	135.3 (15.2)	1.21 (0.15)
4 (b)	139.7 (8.1)	1.31 (0.13)
5 (c)	148.0 (9.6)	1.34 (0.13)
6 (b)	143.5 (14.9)	1.32 (0.23)
7 (a)	194.1 (7.3)	1.78 (0.10)
7 (b)	183.0 (10.8)	1.75 (0.12)
7 (c)	179.6 (6.8)	1.83 (0.15)
HSD	23.8	0.29

sites, and both variables contributed significantly to SFG, with correlation values (Pearson r) of 0.98 and -0.86 respectively.

Although tissue burdens of trace metals (Cr, Ni, Cu, Zn, Cd, Pb) increased during the deployment period, mussels at all sites increased their body weight substantially, and consequently a reduction in overall tissue concentrations was observed. Tissues were not analysed for organic contaminants.

Average dry tissue weight (n = 11) increased from a pre-deployment value of 0.61 g to between 1.21 and 1.83 g during the exposure period (Table 3). In comparison, shell length showed little change, and body condition indices (Table 3) therefore closely matched weight growth. Three distinct groups were evident. Mussels at the reference site were of similar weight, and significantly heavier than mussels from other sites. There were no statistically significant (p > 0.05) differences in dry weight between mussels from any of the outfall exposure sites. Mussels at all field sites, however, exhibited substantial and significant growth with respect to the pre-deployment sample.

3.2 Laboratory Study

All mussels exposed to 8% effluent died within the first week. The clearance rates of mussels in control vessels were consistent throughout the experimental period (1.92-2.19 l h^{-1}, Table 4) and were closely comparable to reference site

Table 4 *Clearance Rates in Mussels Exposed to Effluent Samples in Laboratory over Four Weeks (Average and 95% Confidence Interval)*

Concentration	Week 1	Week 2	Week 3	Week 4
0	1.92 (0.22)	2.03 (0.38)	2.19 (0.31)	2.13 (0.22)
0.08%	2.06 (0.38)	1.83 (0.29)	2.48 (0.45)	2.13 (0.45)
0.15%	2.10 (0.19)	2.17 (0.73)	2.47 (0.25)	2.17 (0.60)
0.25%	1.72 (0.30)	1.26 (0.34)	2.31 (0.34)	2.56 (0.64)
0.45%	0.33 (0.26)	0.90 (0.58)	1.93 (0.46)	2.16 (0.18)
0.80%	0	0.33 (0.10)	1.47 (0.29)	1.78 (0.35)

Table 5 *Clearance Rate in Mussels Exposed to Effluent: % of Control (Approximate Solvent Concentration, mg l^{-1} in Brackets)*

Concentration	Week 1	Week 2	Week 3	Week 4
0	100	100	100	100
0.08%	107 (0.46)	90 (0.33)	113 (0.27)	102 (0.18)
0.15%	109 (0.87)	107 (0.62)	112 (0.51)	101 (0.34)
0.25%	90 (1.45)	62 (1.04)	108 (0.86)	120 (0.57)
0.45%	17 (2.62)	44 (1.87)	88 (1.55)	101 (1.03)
0.80%	0 (4.66)	16 (3.32)	67 (2.76)	83 (1.84)

clearance rates on 16 and 18 August. Clearance rates were substantially reduced at 0.45% and 0.8% effluent in weeks one and two, and also at 0.25% in week one (Table 4).

A progressive reduction in effect on clearance rate was evident over the four-week period at higher concentrations, and no significant effects were apparent at any concentration up to 0.8% by the end of the fourth week. At 0.15% and 0.08% there was some indication of an increase in clearance rate with respect to controls by the end of the third week (Table 5).

At the end of the experimental period, the discharger provided an analysis of the effluent samples for the major consented organic constituents. This indicated a reduction in pH (12.5-8.5) and also a substantial reduction in the sum of consented organic solvent concentrations (583, 416, 345 and 230 mg l^{-1} in successive weeks). Clearance rates at 0.45% and 0.8% were strongly correlated (r = 0.98 and 0.92 respectively) with the sum of consented organic solvent concentrations. Although clearance rates at these concentrations were also correlated with effluent pH, daily monitoring indicated that pH was well-buffered in the experimental tanks.

4 DISCUSSION

The results of the field and laboratory studies demonstrated statistically significant sub-lethal effects of the effluent. In the field, both clearance rate and respiration rate exhibited significant responses. These variables were negatively

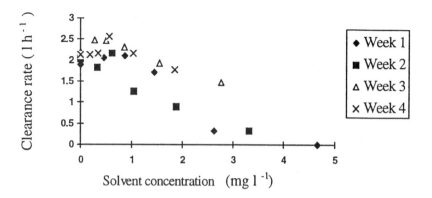

Figure 3 *Relationship Between Solvent Concentration and Clearance Rate*

correlated; a comparison with laboratory results indicated that clearance rate was reduced at higher effluent concentrations, and consequently it is inferred from this that, in the field, elevated respiration rates were associated with higher levels of exposure and stress.

Field SFG studies revealed a significant ($p < 0.05$) reduction at sites one and two (north of the industrial outfall), although SFG at all outfall sites was lower than at the reference sites. SFG in the third control group was significantly higher than in the first two control groups. This was primarily due to higher feeding rate in the third group, but also partly due to relatively lower respiration rates in this group. The laboratory seawater supply was changed between the second and third retrieval dates, and a difference in water quality may have contributed to the observed differences in physiological rates. This observation illustrated the value of repeated reference site measurements, since in the absence of these data an incorrect inference might have been drawn concerning the relatively high SFG value at site five. In the event, it was possible to place the response of mussels from site five in a context comparable to mussels from the other outfall sites.

Laboratory exposure of mussels to effluent demonstrated that there was a quantitative relationship between reduction in feeding rate and the concentration of organic solvents in the effluent (Figure 3). Although a concentration of 8% was lethal, sub-lethal effects were apparent only at concentrations of between 0.25% and 0.8%. The limit of the industrial outfall mixing zone is defined by an effluent concentration of 0.08% (1:1250 dilution) and is presumed to have an approximate radius of 300-400 m. The consent condition operating at the time of the study required that no acute toxicity be evident at ten times the concentration at the edge of the mixing zone. The effluent thus appeared to conform to the consent condition, in that no sub-lethal effect was evident at up to twice the concentration assumed at the edge of the mixing zone. This observation was corroborated by the field study, which showed detectable sub-lethal effects only within 150 m of the outfall; these effects were of a magnitude consistent with a dilution of between 1:200 and 1:400, which is in turn consistent with the distance from the outfall diffuser.

The magnitude of response in the first two weeks of laboratory exposure was sufficient to permit the calculation of approximate EC50 values, using the moving average-angle method.[5] EC50 values were similar, at 2.04 mg l^{-1} and 1.57 mg l^{-1} in weeks one and two respectively. It is notable, however, that these values do not reflect the general trend towards diminishing effect with decreasing total solvent concentration, and it is apparent (Table 5) that there was a substantial change in the shape of the response curve between weeks one and two. Although analytical details of the effluent are not presented here, it is pertinent to observe that only one component did not decline in concentration on all occasions. Methanol concentration increased between weeks one and two, declining thereafter in parallel with the other components.

Mussels placed to detect possible effects from the adjacent sewage discharge did not reveal any marked adverse effects, which suggested that a substantial fraction of the responses observed in mussels from sites one and two could be attributed directly to the effects of the industrial outfall.

The reference site was most clearly distinguished from the outfall in terms of dry tissue weight. This variable did not, however, distinguish between the outfall sites as clearly as did the physiological measurements.

Although laboratory studies indicated a clear relationship between sub-lethal toxicity and reported organic solvent concentration, it was also evident (as a consequence of the change in effluent solvent content) that the observed responses were neither permanent nor persistent. As effluent solvent content decreased, a marked recovery was observed even in those mussels which had exhibited complete inhibition of feeding after the first week of exposure. This is consistent with the nature of the effluent components, which were largely alcohols, and which might be expected to cause reversible non-polar narcosis. Although, in this investigation, post-exposure recovery studies were not conducted on either the field- or laboratory-exposed mussels, such measurements would have provided a valuable indication of the completeness of recovery, and it is recommended that this be taken into account in the design of future studies. While changes in effluent composition exerted a clear influence on laboratory responses, the data provide evidence to advise caution in simple interpretation of effluent EC/LC50 values - changes in concentration-response relationships (e.g. slope) may be associated with alteration in effluent composition without this being clearly reflected in quantitative effects estimates.

Overall, the combination of field and laboratory studies provided a consistent and reasonably sensitive evaluation of the *in situ* effects of an industrial discharge. The laboratory studies contributed an essential calibration step, by demonstrating that one of the primary response endpoints (feeding rate) was able to 'detect' the effluent at environmentally-realistic concentrations. The field studies, in turn, provided an increased level of confidence by allowing the experimental animals to integrate the variation in effluent composition and concentration in the receiving water over an extended period of time. Together, the studies were able to provide confirmation that the existing consent conditions were adequately protective of the receiving waters.

The results of this investigation are instructive, because the consented components of the effluent are neither highly toxic nor highly persistent, but nevertheless exert a measurable (if potentially reversible) effect on surface water quality within the discharge mixing zone. The depressive effect at the population level of chemicals with a general narcotic mode of action may not be well-predicted by laboratory measurements of acute lethal toxicity (except to the extent that such measurements fail to distinguish between mortality and narcotic immobilisation). Donkin et al [6] have demonstrated that consistent quantitative structure-activity relationships can be developed for the effects of hydrophobic chemicals on sub-lethal (and in particular, feeding rate) responses in *Mytilus*.

Although the term SFG has been used, for convenience, throughout this paper, it is recognised that it is only an approximation which permits different physiological rates to be expressed and compared in common energy terms. For the technical reasons outline above, it is not practicable or reliable to measure all the components contributing to 'real' SFG with sufficient precision; consequently, the actual net energy budgets of the animals *in situ* remains an unknown. The objective of this study was not, however, to estimate the energetics effects of discharges, but to evaluate sub-lethal toxicity; in this sense, we are less concerned with factors such as local food quality and absorption efficiency and are more directly concerned with the interaction of contaminants with specific physiological processes. It is conceivable, therefore, that an accurate estimation of actual *in situ* SFG might provide a picture very different from the one presented here. Such a picture would answer a different question:

What is the potential risk to local populations?

rather than:

What is the *in situ* sub-lethal toxicity of the effluent?

The former question may, ultimately, be the most important to answer, since it addresses the issue of risk estimation and management, and recognises that sub-lethal effects may be mitigated in reality by factors such as local variation in food quality. However, it also raises political questions, since a corollary might be that organic enrichment of an effluent could be deemed to compensate for a given level of contaminant loading. The second of the above questions is both more conservative, and provides a more consistent basis for equitable regulation of industrial and sewage effluent.

What remains to be established is the extent to which sub-lethal measurements of the type reported here are adequately protective of receiving waters. There is a degree of tautology in debates concerning the choice of bioassay and toxicity test method, which can make it difficult to determine empirically whether an absence of observed effect in a given situation is due to lack of impact or to lack of sensitivity. There are two aspects to sensitivity. One is the subtlety of the response (i.e. growth versus survival), the other is the exposure level at which the response occurs. The former is required to demonstrate whether or not a particular contaminant has the potential to interfere with critical population processes, while the latter is essential if we are to estimate the potential to cause such effects in specific circumstances. The study reported here indicates that, for the discharge investigated, both requirements were met. It is, nevertheless, important to bear in mind that the study has not in any way demonstrated that decisions based on sub-lethal responses in mussels will be directly or fully protective of the receiving waters. The application of *in situ* sub-lethal effects monitoring represents a step forward, by removing a layer of uncertainty associated with acute lethal laboratory studies; it does not represent a final answer. The essential step in adopting any biological measurement method is not to 'prove' that it is valid, but to identify and understand its limitations - that is, a method may not do all that we wish, but we should attempt to determine clearly what it can *not* do. The data generated by this study suggest that, at the least, physiological responses in *Mytilus* permit some discrimination in terms of biological water quality between sites within a relatively small geographical area, and that this approach is capable of 'detecting' effluent at fairly high dilutions. This observation is reinforced by the findings of Butler *et al*,[7] in a study of the sub-lethal effects of sewage sludge on *Mytilus*, that feeding rate was reduced by 50% at a sludge-in-seawater concentration of 0.02%.

References

1. L. W. Hall Jr, M. C. Ziegenfuss, S. A. Fischer (1992). Ambient Toxicity Testing in the Chesapeake Bay Watershed Using Freshwater and Estuarine Water Column Tests. *Environmental Toxicology and Chemistry* **11** 1409-1425
2. B. D. Roddie, T. Kedwards and M. Crane (1992). Potential Impact of Watercress Farm Discharges on the Freshwater Amphipod, *Gammarus pulex* L. *Bulletin of Environmental Contamination and Toxicology* **48** 63-69
3. B. D. Roddie and Butler (1989). Sewage Discharge to Sea and Stress Indices: Final Report to the Department of the Environment. DoE 2147-M, 31 pp
4. B. L. Bayne, D. A. Brown, K. Burns, D. R. Dixon, A. Ivanovici, D. R. Livingstone, D. M. Lowe, M. N. Moore, A. R. D. Stebbing and J. Widdows (1985). The Effects of Stress and Pollution on Marine Animals. Praeger, 386 pp
5. US EPA (1993). Short-term Methods for Estimating the Chronic Toxicity of Effluents and Receiving Waters to Marine and Estuarine Organisms. Second Edition. EPA-600/4-91/021
6. P. Donkin, J. Widdows, S. V. Evans, C. M. Worrall, and M. Carr M (1989). Quantitative Structure-activity Relationships for the Effects of Chemicals on the Rate of Feeding by Mussels (*Mytilus edulis*). *Aquatic Toxicology* **14** 227-294
7. R. Butler, B. D. Roddie and C. P. Mainstone (1990). The Effects of Sewage Sludge on Two Life History Stages of *Mytilus edulis*. *Chemistry and Ecology* **4** 211-219

Sub-lethal Effects of Waste on Marine Organisms: Responses Measured at the Whole Organism Level

P. Donkin,[1] J. Widdows,[1] D. M. Lowe,[1] M. E. Donkin[2] and D. N. Price[2]

[1]Plymouth Marine Laboratory, Prospect Place, West Hoe, Plymouth PL1 3DH, UK
[2]Plymouth Environmental Research Centre, University of Plymouth, Drake Circus, Plymouth, PL4 8AA.

ABSTRACT

The marine environment is a sink for a diverse mixture of waste chemicals arising from direct inputs through pipelines or from ships, from land run-off to rivers, storm drains and sewers and from atmospheric deposition. One of the biggest challenges of marine aquatic toxicology is to determine the extent to which the combined effects of chemical waste from many sources adversely impacts the biota. Considerable progress has been made recently in developing sub-lethal effects measurements which may be used to determine the health of organisms taken from specific environmental locations. However, since wastes include chemicals with many modes of toxic action, from non-specific narcotic toxicants such as hydrocarbons to pesticides designed specifically to be neurotoxic and phytotoxic, no single bioassay can meet the needs of *in situ* environmental impact assessment. In this paper we describe field and laboratory studies with the mussel *Mytilus edulis* and with species of macroalgae.

The physiologically determined energy balance (Scope for Growth: SFG) of mussels has been used extensively to monitor the impact of organic contaminants. Cause-effect relationships have been established by measuring contaminant body burdens and comparing these with dose-response relationships established in the laboratory. The usefulness of these data have been extended by establishing QSARs. The presence of some pesticides can also be indicated by the activity of specific enzymes (e.g. acetylcholinesterase). The presence of unsuspected toxicants can be demonstrated by tissue extraction/fractionation/toxicity assay experiments.

Macroalgae, like mussels, are common and readily harvested for chemical analysis and physiological study. They are particularly susceptible to herbicides, which often act by inhibiting photosynthesis. Methods for demonstrating the impact of specific and non-specific toxicants on macroalgae are described.

1 INTRODUCTION

Any anthropogenic chemical which is no longer functioning in its intended role may be described as waste. Many such wastes enter the aquatic environment through controlled industrial and urban wastewater discharges but there are also substantial inputs from diffuse sources such as atmospheric fallout and land runoff. There is an increasing trend towards monitoring and controlling the impact of wastewater inputs by means of bioassays. This approach allows even the most complex and chemically ill-defined effluents to be regulated in an environmentally meaningful way.

Although such trends are a worthwhile advance, the impact of human activity on the environment is the consequence of the combined effects of all our wastes both from individual and diffuse sources. Material leaving a pipeline may be harmlessly dispersed in the aquatic environment or may be concentrated through partitioning processes and combine with effluents from elsewhere to form a hazardous mixture. If we are to protect the environment from damage, and remediate damage already done, it is essential that we develop a capacity to measure total human impact.

Considerable progress has been made in recent years in developing sub-lethal effects measurements which can be applied to organisms taken from the field. Because of the complexity of chemical wastes and the number of different modes of toxic action by which they act, no single bioassay can meet all monitoring requirements. In this paper we will discuss studies in which the marine mussel *Mytilus edulis* and various species of macroalgae have been used successfully.

2 SUB-LETHAL EFFECTS MEASUREMENTS USING MUSSELS (*MYTILUS EDULIS*)

Mussels and related bivalve molluscs have been used extensively throughout the world in environmental monitoring programmes. Measurements of sub-lethal biological effects and measurements of tissue residues of chemical contaminants have both been included in such programmes.[1] One of the most successful sub-lethal effects measurements applied has been the determination of the energy balance (termed Scope for Growth; SFG) by means of whole-animal physiological measurements. In the vicinity of a single major source of contamination such as an oil terminal, close links between reduced SFG and tissue residues of contaminants are easily established and the identity of the source is uncontroversial.[2] However, studies carried out along a pollution gradient in Bermuda showed high (p > 0.75) inverse correlations between SFG and tissue residues of lead, PCBs, hydrocarbons, polar hydrocarbon oxidation products and tributyltin.[3] It was impossible to tell from this study whether an individual compound or compound group was causing the impact, or whether it was the consequence of the combined effect of them all. The problem of covariance between contaminants is reduced in studies over a larger spatial scale, such as our investigation of mussel populations along the UK North Sea coast,[4] but the task of relating observed effects to environmental factors is an even greater challenge. In this study, hydrocarbons derived from urban/industrial activity appeared to be the most important single toxicant class, but much of the observed biological impact could not be explained by our existing understanding of tissue burden-response relationships.

We have sought to improve our capacity to link effects to contaminants in four ways.

1   To establish body burden-response relationships in the laboratory for compounds which have unique modes of action.
2   To link whole animal physiological responses to relevant biochemical biomarkers.
3   To establish quantitative structure-activity relationships (QSAR) on both an aqueous and tissue concentration basis. QSARs provide a rational basis for extrapolation of limited laboratory data to untested compounds detected in the environment, assist in establishing the mode of action of compounds on the biological endpoint chosen for bioassay, and facilitate prediction of the likely impact of complex mixtures.
4   To extract organic contaminants from mussels known to be stressed, fractionate the extracts, test their toxicity, then identify the chemicals in the toxic fractions.

The antifouling compound tributyltin is one of the most important toxicants in the marine environment possessing unique toxicological properties. It adversely influences SFG at low concentrations by a combination of respiratory uncoupling and direct action on feeding rate. Body burden-response relationships have been determined which can assist in interpreting field data.[5,6]

Acetylcholinesterase (AChE) inhibiting pesticides (organophosphates and carbamates) have been shown to be toxic to the feeding activity of mussels (quantitatively the most important and sensitive SFG parameter) and this inhibition is associated with suppression of AChE activity in the mussels' gills.[7] Although compounds other than established neurotoxins have been shown to inhibit AChE activity in some organisms,[8] alteration of physiological and behavioural parameters in combination with changes in enzyme activity may provide strong diagnostic evidence for impact caused by these pesticides.[9,10,11]

QSARs which have been established for the effect of hydrocarbons on the

Table 1 *The Effect of Pesticides on the Feeding Rate of* Mytilus edulis: *Comparison of Observed and Predicted Toxicity Expressed as Concentration in* M. edulis *Tissue Required to Reduce Feeding Rate by 50%, (TEC50).*

Compound	Log K_{OW}	TEC50 Obs. (μmol kg^{-1} wet wt.)	EC50 Narc.[a] /EC50 Obs.
Pentachlorophenol	5.12	66	3.4
2,4,5-Trichlorophenol	3.72	201	1.1
2,4-Dichlorophenol	3.08	982	0.3
Tributyltin	3.80	5	46.7
Dibutyltin	1.49	39	5.8
Di(2-ethylhexyl)phthalate	5.11	170	1.32
Lindane	3.61	1030	0.2
Endrin	4.56	260	0.9
Carbaryl (Sevin)	2.36	50	4.5
Dichlorvos	1.16	2	112.0

[a] TEC50 narcosis = 224 ± 89 (std. dev., n = 14) μmol kg^{-1} wet. wt.

feeding rate of mussels[12] suggest that these compounds act as non-specific narcotic toxicants. Body burden measurements demonstrated that many hydrocarbons reduced feeding rate to 50% of control values at approximately the same tissue concentration (224 ± 89 (std. dev., n = 14) μmol kg^{-1} wet wt.). This value can be used as a baseline to compare the toxicological properties of other compounds studied in the laboratory. Such a comparison (Table 1) showed that tributyl- and dibutyltin, the respiratory uncoupler pentachlorophenol and the AChE inhibiting pesticides, dichlorvos and carbaryl, exhibit enhanced toxicity towards mussel feeding activity over that expected of non-specific narcotics.[13] However, the organochlorine pesticides, lindane and endrin, show no elevation of toxicity. Similarly, low toxicity was observed for pyrethroid pesticides. The organochlorine and pyrethroid pesticides are potent neurotoxins in their target organisms and are also extremely toxic towards aquatic crustaceans.[7,13] Thus, although the mussel is an excellent indicator organism for the impact of narcotic chemicals on the marine environment, and is also a useful indicator of the effects of respiratory uncouplers and AChE inhibiting pesticides, it tells us little about the likely impact of organochlorine and pyrethroid neurotoxins. Crustaceans are better monitoring organisms for this purpose.

Linking effects to chemical causes described above depend substantially upon knowing and being able to measure the contaminants present. This is not always feasible because of the complexity of many of the contaminant mixtures which enter the environment. We have attempted to address this problem using a fractionation / toxicity testing approach.[14] Organic contaminants were extracted from the tissues of mussels known to exhibit reduced SFG, the extracts fractionated, then the fractions tested using two assays known to be linked to the poor performance of mussels in the field. These assays were feeding rate, modified by the use of small mussels to allow assay of small

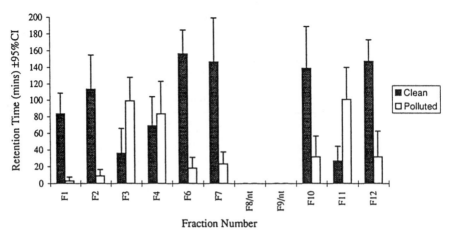

Figure 1 *Neutral Red Retention in the Lysosomes of Blood Cells from* Mytilus Edulis: *Cells Exposed to Fractionated* M. edulis *Tissue Extracts from a Clean and a Polluted Site. (Mean ± 95% C.I.)*

volumes of water, and the retention of neutral red dye in the lysosomes of blood cells.[15]

The results from a neutral red assay of fractions obtained from a low stress (relatively clean) and high stress (contaminated) site are shown in Figure 1. Similar results were obtained for samples collected at the same site the previous year though the feeding rate data (not shown) were less reproducible. Toxic biochemical constituents of the mussels influence the feeding rate and neutral red data for some fractions. However, data from the remaining fractions indicated that the SFG of mussels from the polluted site was probably reduced by a complex mixture of bioaccumulated organic contaminants, many of which are as yet unidentified. Though considerable development work is still required to improve the reproducibility of this procedure, it has potential for highlighting the presence of toxic contaminants not determined in routine chemical monitoring programmes.

3 SUB-LETHAL EFFECTS MEASUREMENTS USING MACROALGAE

Macroalgae have great potential as monitoring organisms since they, like mussels, occupy a fixed location and are large enough to facilitate easy chemical analysis. They are particularly sensitive to herbicides which are quantitatively important organic contaminants of aquatic, including estuarine, environments.[16]

We have developed QSARs for the effect of non-specific narcotic alcohols on two parameters, whole frond neutral red retention and ion leakage, in the green macroalga *Enteromorpha intestinalis*.[17] Ion leakage, or the "health index" calculated from it, has proved to be best suited to field studies and has

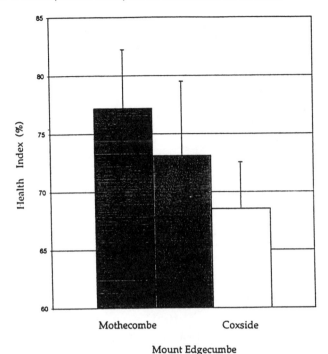

Figure 2 Ion Leakage Health Index of Fucus vesiculosus from Three Sites near Plymouth : Mothecombe, Clean; Mount Edgecumbe, Slight Pollution; Coxside, Highly Polluted. (Mean ± std. dev.).

demonstrated differences in the condition of macroalgae taken from clean and polluted sites; Figure 2 presents data for the brown macroalga *Fucus vesiculosus*. Ion leakage is a good indicator of non-specific effects, but most herbicides have specific modes of action, frequently involving inhibition of photosynthesis.[18] Some herbicides induce kinetic changes in the fluorescence output of illuminated plants which reflect reductions in photosynthetic efficiency.[19] Recently, portable fluorescence meters designed specifically to measure these changes in the field have become available. A concentration response relationship for the effect of the antifouling triazine herbicide irgarol on the fluorescence parameter F_v/F_m ratio in *E. intestinalis* is shown in Figure 3. F_v is the variable fluorescence and F_m the maximum fluorescence. Irgarol levels equivalent to the EC50 have been detected in Mediterranean marinas[20] so this method clearly has potential for the rapid determination of herbicide impact in the environment.

4 CONCLUSIONS

Sub-lethal effects measurements currently available are capable of detecting the impact of environmental contamination in aquatic environments remote

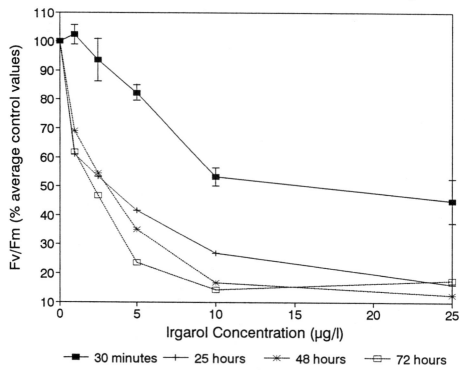

Figure 3 *The Effect of Irgarol on the Fluorescence Induction Ratio (Fv/Fm) in Enteromorpha intestinalis. (For Clarity, Errors, Mean ± s.e., Are Shown on One Line Only).*

from individual wastewater inputs. An understanding of the selective toxicity of organic contaminants towards different biological endpoints is essential for meaningful interpretation of such measurements. Linking chemical causes to observed effects remains a challenge, but progress is being made.

ACKNOWLEDGEMENTS

The work described forms part of the research programme of the Plymouth Marine Laboratory (a component of the Natural Environment Research Council) and the University of Plymouth. The authors are also grateful for the support of the National Rivers Authority, the (UK) Department of the Environment and the Polytechnics Funding Council. Sarah Dashfield (University of Plymouth) produced the ion leakage data.

References

1 J. Widdows and P. Donkin, In: "The Mussel Mytilus: Ecology, Physiology, Genetics and Culture," (ed. E. Gosling), Elsevier press, Amsterdam, 1992, Chapt. **8**, 383.

2 J. Widdows, P. Donkin, S. V. Evans, D. S. Page, and P. N. Salkeld, *Proc. Roy. Soc. Edinb.*, 1995 **103B**, 99.
3 J. Widdows, K. A. Burns, N. R. Menon, D. S. Page and S. Soria, *J. Exp. Mar. Biol. Ecol.*, 1990, **138**, 99.
4 J. Widdows, P. Donkin, M. D. Brinsley, S. V. Evans, P. N. Salkeld, A. Franklin, R. J. Law and M. J. Waldock, *Mar. Ecol. Prog. Ser.*, 1995, **127**, 131.
5 J. Widdows, and D. S. Page, *Mar. Environ. Res.*, 1993, 35, 233.
6 J. Widdows, and P. Donkin, *Comp. Biochem. Physiol.*, 1991, **100C**, 69.
7 P. Donkin, J. Widdows, S. V. Evans, F. J. Staff and T. Yan, submitted, *Pesticide Sci.*
8 M. Nendza, A. Wenzel and G. Wienen, *SAR QSAR Environ. Res.*, 1995, **4**, 39.
9 J. G. McHenery, G. E. Linley-Adams and D. C. Moore, *Scottish Fisheries Working Paper No. 16/91*, The Scottish Office Agriculture and Fisheries Department, Aberdeen, 1991.
10 J. G. McHenery and S. W. Forsyth, *Scottish Fisheries Working Paper No. 8/91*, The Scottish Office Agriculture and Fisheries Department, Aberdeen, 1991.
11 R. Serrano, F. Hernández, J. B. Peña, V. Dosta and J. Canales, *Arch. Environ. Contam. Toxicol.*, 1995, **29**, 284.
12 P. Donkin, J. Widdows, S. V. Evans, and M. D. Brinsley, *Sci. Total Environ.*, 1991, **109/110**, 461.
13 P. Donkin, J. Widdows, S. V. Evans, and F. J. Staff, Prediction of effects of organic contaminants on mussels. Foundation for Water Research report FR/D 0016, Marlow, UK, 1994.
14 A. Castaño, M. Vega, T. Blazquez, and J. V. Tarazona, *Environ. Toxicol. Chem.*, 1994, **13**, 1607.
15 D. M. Lowe, V. U. Fossato and M. H. Depledge, *Mar. Ecol. Prog. Ser.*, 1995, **129**, 189.
16 J. L Zhou, T. W. Fileman, S. V. Evans, P. Donkin, R.F.C. Mantoura and S.J. Rowland, *Mar. Pollut. Bull.*, 1996, (in press).
17 R. Schild, P. Donkin, M. E. Donkin and D. N. Price, *SAR QSAR Environ. Res.*, 1995, **4**, 147.
18 S. O. Duke, *Environ. Hlth. Persp.*, 1990, **87**, 263.
19 M. Grouselle, T. Grollier, A. Feurtet-Mazel, F. Ribeyre and A. Boudou, *Ecotoxicol. Environ. Saf.*, 1995, **32**, 254.
20 Readman, L. L. W. Kwong, D. Grondin, J. Bartocci, J. P. Villeneuve, and L. D. Mee. *Environ. Sci. Technol.*. 1993, **27**, 1940.

The Toxic Impact of Petro-chemical and Oil-refining Waste on Hydrobionts

N. G. Kuramshina, E. M. Kuramshin and S. V. Pavlov

The Institute for Problems of Applied Ecology and Natural Resources Use, 12/1, 8 Marta Str., 450005, Ufa, Bashkortostan, Russia

ABSTRACT

This paper describes research carried out on the potential impacts of petro-chemicals and oil refining wastes on molluscs. The acute and chronic effects of 4,4-dimethyl-1,3-dioxane on gastropod molluscs are described. Studies were also conducted on the effects of diesel fuel, aviation fuel and multigrade oil.

The results indicate the importance of assessing both lethal and sub-lethal effects. Sub-lethal effects noted included behavioural changes and effects on hemolymph.

1 INTRODUCTION

Due to the rapid increase in surface water pollution by the petro-chemical and oil-refining industry, the maintenance of water purity by means of biological regulation has become one of the most important problems of society. At present, bio-testing with hydrobionts is being used to evaluate the toxicity of chemicals discharged with industrial waste waters. Molluscs, widely spread hydrobionts, are playing an important role in the chain of ecological relations and biological purification of waterbodies. Chemicals getting into water cause abnormality of metabolism, breathing activity, and fertility in molluscs. All these factors allow the use of molluscs as bio-indicators of aquatic environmental quality.

2 METHODS

Research was carried out on freshwater gastropods, *Anisus vortex* and *Planorbis rubrum,* under laboratory conditions in aquaria with sample groups of animals including 20–25 individuals at a temperature of 21 °C. The highly water soluble material, 4,4-dimethyl-1,3-dioxane (DN), which is a by-product of the large-scale production of isoprene, was considered as a toxicant of the petrochemical industry.

Acute toxicity after 48 hours exposure was defined by behavioural and physiological criteria. Chronic toxicity was evaluated after 60 days through changes in morphological parameters (shell size, mollusc mass). The impact of toxicants on the physiological state of molluscs was studied by the functional state of blood cells which had been stained with a vital dye.

3 RESULTS

After being exposed to DN for 48 hours, molluscs were observed to undergo behavioural changes, evidenced by body contraction, pulling in of tentacles, getting inside the shell, cessation of movement, and lateral position. Under these conditions, lethal (100%), median effective (50%) and bearable by molluscs (threshold) concentrations of DN in water were 0.05; 0.01 and 0.001 mg l^{-1}, respectively. A rise in temperature from 15 to 25 °C resulted in toxicological effects occurring some two to six times more rapidly, depending on DN concentration.

Chronic experiments (60 days) at a DN concentration of 0.001 mg l^{-1} (tolerable for molluscs) revealed an adverse impact on the morphometric parameters of molluscs. The observed suppression of shell diameter increase and of total mollusc mass reached 30% as compared with the control group of animals. Prolonged exposure to DN caused changes in the functional state of blood cells. Abnormality of haemocyte membrane permeability was observed, the percentage of non-viable cells reaching 50%. These results confirm high toxicity of 4,4-dimethyl-1,3-dioxane, which causes morpho-physiological changes in molluscs even at a concentration (0.001 mg l^{-1}) lower than the maximum permissible concentration (0.005 mg l^{-1}) stated by sanitary - toxicological indicator for water discharge.

Diesel fuel (DF), aviation fuel (AF) and multigrade oil (MO) were used as toxic products of oil refining. Maximal concentrations of toxicants were derived as a result of studying behavioural reactions of the mollusc *Planorbis rubrum* after 48 hours exposure. The concentrations for DF, AF and MO were 0.1, 0.1 and 5 ml l^{-1} respectively. Mollusc hemolymph was used for evaluation of the chronic toxic impact of fuels and oils as it appears to be one of the more sensitive physiological systems.

In an evaluation of DF toxicity (0.1 ml l^{-1}) in chronic experiments (30 days), an insignificant increase in the relative content of mollusc hemolymph (from 0.48 to 0.52) and a decrease in the number of cells (by approximately four times) were observed as compared with the control group of animals. Mollusc mortality was 20%.

The impact of aviation fuel (0.1 ml l^{-1}) on molluscs in chronic experiments (30 days) manifested itself by almost a doubling in the number of blood cells.

The relative content of mollusc hemolymph as a result of aviation fuel impact remained at control levels. Mollusc mortality was 20%.

There was no clear effect of multigrade oil on molluscs. However, there were some insignificant changes in the relative content of mollusc hemolymph and in blood cell numbers as compared with the control group. In the control group of molluscs the relative content of hemolymph was 0.48, the number of hemolymph cells in 1 cm^3 reached 2.5×10^2.

4 CONCLUSION

Studies of the impact of various toxicants on freshwater gastropods have shown that for evaluation of threshold concentrations it is important to combine the study of the behavioural reactions of molluscs with qualitative and quantitative analysis of mollusc hemolymph.

The Role of Environmental Quality Standards in Controlling Chemical Contaminants in the Environment

N. G. Cartwright and S. Lewis

National Centre for Environmental Toxicology, WRc, Henley Road, Medmenham, Marlow, Bucks. SL7 2HD, UK

ABSTRACT

The Environmental Quality Objective (EQO) / Environmental Quality Standard (EQS) approach has long been the cornerstone of the UK approach to controlling the release of dangerous substances to the aquatic environment.

The National Centre for Environmental Toxicology at WRc plays a leading role in this area by deriving EQSs for the Department of Environment and for the Environment Agencies in England and Wales, Scotland and Northern Ireland. EQSs exist for over 60 substances and a further 30-40 are currently being derived in on-going research programmes.

This paper discusses EQS derivation, the basis of the EQS approach and how new approaches to pollution control may impact on this in the future.

1 INTRODUCTION

Three main approaches are adopted in the EU to control chemical contaminants in the environment:

1. Product control (risk assessment of exposure and environmental effects from use of the product, prior to its registration);
2. Technological limits (using Best Available Technology); and
3. Environmental Quality Objectives/Standards (for individual substances)

A fourth, used to a lesser extent, is the application of economic instruments (e.g. levying charges based on emission levels).

This paper discusses the derivation of Environmental Quality Standards and how their role in pollution control may change in the future.

2 BASIS OF ENVIRONMENTAL QUALITY STANDARDS (EQSs)

The EC Dangerous Substances Directive (76/464/EEC)[1] and Daughter Directives establish EQSs for the most harmful (List I) substances and require individual Member States to set EQSs for less harmful List II substances. The National Centre for Environmental Toxicology, WRc, has ongoing research programmes with the UK Department of the Environment (DoE) and Environment Agencies to derive EQSs. Standards proposed are peer-reviewed by an external scientific committee and, those derived for the DoE, go to public consultation before being incorporated into Statutory Instruments.

The derivation of Environmental Quality Standards presupposes that there will be a concentration of a substance in the aquatic environment at which the intended quality objective (e.g. protection of the aquatic ecosystem) will not be compromised and that this can be quantified.

In practice, however, this is rarely straightforward. In particular, there are two problem areas. The first of these is defining the quality objective precisely. The second is identifying suitable endpoints and obtaining sufficient data to enable standards to be set which will meet the required objective without being over-protective. Defining the objective in sufficiently precise terms to enable selection of appropriate endpoints is particularly difficult. Unlike human risk assessment, where the aim is to protect individuals in a population, the aim of EQSs is to protect the viability of an aquatic ecosystem in the short and long-term.

The EC Scientific Advisory Committee for Toxicology and Ecotoxicity of Chemicals[2] suggest that Water Quality Objectives (equivalent to EQSs) set for List I substances should:

1   permit successful completion of all stages in the life of aquatic organisms;
2   not produce conditions which cause these organisms to avoid parts of the habitat where they would normally be present;
3   not give rise to accumulation of substances that can be harmful to biota (including man); and
4   not produce conditions which alter the functioning of the ecosystem.

The UK (like Canada) aims to achieve adequate protection of the aquatic environment by protecting the most sensitive species and life-stages by application of empirical safety factors to the lowest credible adverse effect concentrations. The Dutch, in comparison, use extrapolation procedures to identify a hazardous concentration for 5% of species (Maximum Tolerable Concentration) with 50% confidence. This implies that adequate protection of the aquatic environment will be achieved by protecting 95% of species.

3 DATA REQUIREMENTS

The calculation of an Environmental Quality Standard is primarily based on an assessment of the available toxicity data on aquatic organisms, from laboratory studies and, where available, field observations. It also requires an understanding of the physico-chemical properties (e.g. solubility, partitioning), the environmental fate and behaviour (e.g. persistence: biotic and abiotic degradation, volatilisation), and the bioaccumulation potential of the substance.

3.1 Physico-Chemical Properties

It is important to have reliable values for those physico-chemical parameters which play a particularly important role in a chemical's fate in the environment, especially in water. These would include, for example: solubility; volatility (Henry's Law constant); and octanol-water partition coefficient (K_{OW}).

It is also necessary to know how the chemical is likely to behave under experimental conditions so that results from laboratory toxicity studies can be properly assessed. In particular, chemicals with low solubility, high volatility or high sorption potential may present problems associated with preparation of solutions or maintenance of appropriate exposure concentrations, and test design should compensate for this.

3.2 Environmental Fate and Behaviour

The physico-chemical data can only give a broad idea of the likely behaviour of a chemical in the environment. Information on the speciation of a chemical in natural waters, the degree to which it sorbs to particulates or sediment, how persistent it is and if it degrades in the environment, is important in assessing how bioavailable a substance is likely to be. In addition, these data indicate whether degradation products should be considered and provide data which assist in deciding relevant durations of exposure for toxicity testing.

3.3 Bioaccumulation

Results from bioaccumulation (from surrounding media) studies can often be predicted from simple physico-chemical measurements such as K_{OW} values. However, at best, such predictions can only give an idea of the order of magnitude of bioaccumulation, and do not take account of biomagnification (from food). Experimental study of these processes enables a consideration of body burdens to be made, and thus allows predictions of possible long-term deleterious effects. This is particularly important if no chronic toxicity data are available.

A commonly used measurement of a substance's bioaccumulation potential is the bioconcentration factor (BCF), where, the

$$BCF = \frac{\text{concentration in organism at steady}}{\text{concentration in water}}$$

Experimentally derived BCFs are generally considered to be the most reliable[3] although some writers suggest that experimental BCFs are often of questionable quality and that theoretical BCFs calculated from the K_{OW} should be used instead.[4] Mackay[5] for example, gives the relationship BCF = K_{OW} x 0.05 to predict BCFs for fish, with substances which possess a log K_{OW} between 2 and 6. This approach has been recommended by the Organisation for Economic Cooperation and Development (OECD)[3] in the absence of experimental BCFs.

3.4 Aquatic Toxicology

The sensitivities of different species exposed to pollutants can vary widely, both within and between species and it is important that a comprehensive review is undertaken to identify toxicity data for different families and trophic levels. This is particularly important for biocidal substances where the data-set must account for inter- and intra-species variability and, most importantly, ensure that species biologically proximate to the target species have been included. For example, the derivation of a water quality standard for an insecticide would require reliable toxicity data on aquatic insects.

Early life-stages of an organism will often be more sensitive to a particular toxicant and the population effects resulting from impact on these life-stages more severe. Greater confidences can therefore usually be placed in data-sets which include a variety of life-stages.

The effects of pollutants on aquatic organisms can often be considerably different for periods of short-term (acute) exposure and long-term (chronic) exposure. In fact, aquatic organisms are generally more sensitive to chronic exposure than acute exposure. In the environment, organisms may be subjected to short-term episodic pollution or long-term continuous pollution depending on the use and entry of a chemical into the aquatic environment and ideally the data-set used should reflect likely exposure scenarios.

In the Netherlands, a statistical methodology has been adopted which aims to calculate the hazardous concentration for 5% of species (HC5) (or the concentration of a substance at which 95% of species are protected) based on all the available No Observed Effects Concentrations (NOECs).[6]

$HC5 = \exp(xm)/T; T = \exp^{(sm*k)}$

Where xm is the average of logarithmically (ln) transformed NOECs, sm is the standard deviation of logarithmically (ln) transformed NOECs, k is a factor determined from sample size and the desired confidence level (usually 50%), and T is the safety factor to be employed.

The OECD[3] highlighted some of the assumptions on which this approach is based and Smith and Cairns[7] discussed a number of statistical and ecological concerns about the use of such models to establish water quality standards. The statistical models are based on 'tolerance theory', developed to establish engineering tolerances. Smith and Cairns argued that whilst engineering parts can be randomly selected and have no interrelationships, the same is not true for the environment where interrelationships are critical to the functioning of the ecosystem. They were concerned that statistical rigour could lead to a false sense of confidence in the method and did not believe that the assumptions of the models would be sufficiently well met to result in accurate predictions of safety. Some of their concerns related specifically to the statistical assumptions in the models e.g. that LC50s/NOECs for a species in a population could be viewed as a sample from a distribution, effectively ignoring individual sensitivities and variances. However, biological concerns such as whether laboratory organisms exhibit the same range of tolerances as field organisms, the use of LC50s/NOECs as end-points and consideration of confounding factors (e.g. bioavailability, data quality), are issues which have to be addressed whichever approach is used to derive EQSs.

These two papers provide a useful insight into the assumptions behind some of the models and the dangers of applying them blindly without sufficient understanding of their assumption and limitations. They also highlight the importance of scientific judgement in the selection of reliable and relevant data, whether for use in the models or other approaches, and the need for a transparent approach.

An alternative approach to models is the use of empirical safety factors which are applied to the lowest credible toxicity value (e.g. UK, CSTE, EEC, OECD). Most methodologies for deriving water quality standards use safety factors ranging from 1, 10, 100, and occasionally, 1000. The choice of factor depends on the data on which it is being applied, typically in the UK:

1 100 is applied to an acute lethal effects level;
2 10 to a chronic, sub-lethal effects level; and
3 1 (occasionally) to reliable chronic effects value from field studies in natural water bodies.

Some organisations (e.g. CSTE/EEC*, US EPA, OECD) apply 1000 to acute effects levels when few data are available and the uncertainty is therefore greater.

If the toxicity data are reliable and relevant, there is less uncertainty associated with identifying sensitive taxa and determining a lowest effect level. Thus chronic data are more useful than acute data, particularly when attempting to protect against long-term exposure (as is the case for most chemical contaminants), and whenever possible are given preference over acute data when deriving a standard. Unfortunately, many of the data that are available in the literature are derived from acute laboratory studies employing standard test species, which are not necessarily the most sensitive or ecologically relevant species. For example, Sloof et al[8] examined the sensitivity of 22 freshwater species to a total of 15 pesticides, inorganic and organic chemicals. These were compared against a suite of organisms (*Selenastrum capricornutum, Daphnia magna, Poecilia reticulata*) and the ratio of the lowest no-observed

---

*CSTE/EEC in the EC Scientific Advisory Committee on the Exotoxicity of Chemicals (Comité Scientifique Consaltif pour l'examen de la Toxicité et de l'écotoxicité des substances chemiques).

lethal concentration for this suite of species compared to that for the most sensitive of the 22 species was less than 10 for 66% of species, between 10 and 32 for 28% and between 32 and 100 for 6%.

The safety factors described above are largely based on a study by the US EPA[9] which compared acute and chronic data for otherwise similar studies (same species, same test conditions) over a wide range of substances. The US EPA found that for 95% of substances tested, the acute:chronic ratio was within about 2 orders of magnitude, while for 50% of substances the acute:chronic ratio was within 1 order of magnitude. A factor of 10 applied to the lowest acute value should therefore give a reasonable prediction of the lowest chronic effect or no-effect level for an 'average' substance. An additional factor of 10 is generally considered necessary to derive a 'safe' level from a chronic effect or no-effect level, to allow for uncertainty in the identification of the most sensitive species. The US EPA[9] compared acute LC50s of a number of species for different chemicals and concluded that a factor between 2.4 and 48.9 applied to a single 'typical' LC50 would encompass 50–95% of the lower LC50s for that chemical.

However, these assumptions should be treated with caution. For example, while these may hold true for chemicals with a non-specific mode of toxic action, chemicals with specific modes of action are likely to exhibit greater ranges of sensitivities both between species and for different exposure durations. Scientific judgement will be important in selecting appropriate safety factors in these instances. Recent work by ECETOC[10] suggested that for 19 organic chemicals, acute to chronic ratios were in the range 1.25–28.3 but a greater range of ratios was found for other types of substances.

The OECD[11] have compared the two extrapolation approaches, statistical models and safety factors. In general, those methodologies for deriving water quality standards which use a statistical approach demand a large, reliable data-set, whereas those methodologies which utilise safety factors may still be used to derive water quality standards providing data for the target species are available. When using the same data-set, the safety factor approach was equally or more protective.

The safety factor approach has some advantages over the use of statistical models. It is simple and although it does not use all the effect data available, it does allow scientific judgement in the selection of key data (e.g. indigenous species, sensitive life stages) which are of particular relevance to the chemical under consideration. The disadvantage is that when there is any deviation from the standard empirical safety factors, there is no firm basis for adjustment and this must be based on the judgement of competent individuals.

To check the extrapolation process, it is prudent to compare the derived water quality standard with reported field observations, where available. Ideally, all water quality standards would be based on field studies, as these generally give the best indication of true environmental responses to pollutants. However, in practice, field data are rarely available. It is important that the derivation process is transparent to enable recommendations and revision as necessary when new data become available.

4 HOW MIGHT THE ROLE OF EQSs DEVELOP IN THE FUTURE

One of the prime aims of EC environmental policy is sustainable development which implies the achievement of economic growth whilst providing adequate environmental protection.

EQSs establish concentrations of individual substances which should not be exceeded if particular uses are not to be compromised. This approach offers regulators considerable flexibility in application. They determine the uses which a receiving water must be able to meet and can define the size of a mixing zone where the EQS may be exceeded when setting discharge consents. The EQS approach takes into account the pollution load already present in the receiving water from point and diffuse sources and is most appropriate for controlling discharges to receiving waters with a small dilution capacity. It is economically efficient because it only requires reduction of discharges to a level sufficient to protect the specified uses. It is less appropriate, however, for large international waters with large dilution capacities.

With the creation of Environment Agencies in England, Wales, Scotland and Northern Ireland in April 1996 an integrated approach to environmental protection is likely to be emphasised. Integrated Pollution Control (IPC) is already applied to the most polluting industries and it is likely that these will in future be covered by the EC Directive on Integrated Pollution Prevention and Control, whilst a revised Dangerous Substances Directive will cover the discharge of dangerous substances from other sources.

Thus, whilst there is likely to be increased emphasis on the application of clean technology and the Best Available Technology for reducing emissions, EQSs are likely to remain an important control measure. Indeed, the tendency for integrated control measures highlights the need for a consistent approach to setting environmental standards for other media (such as sediments and the terrestrial environment) to enable selection of the Best Practicable Environmental Option. The Framework Directive currently being drafted by the EC could result in the repeal of other use-related Directives (such as those for Groundwater, Freshwater Fish and Shellfish). Amongst other requirements, this is likely to lead to identification of operational targets for ecological quality and establishment of a more comprehensive monitoring system for water quality. The Environment Agency is currently developing a General Quality Assessment Scheme which would enable a more integrated assessment of physical, chemical and biological parameters. We can thus expect more integration between ecological monitoring, ecotoxicological assessment and chemical standards in the future to enable effective prioritisation of operational activities within a catchment planning framework and cost effective monitoring.

Although the combination of technology standards and EQSs have been used in IPC to reduce emissions from the most polluting industries, there is concern that single chemical standards do not adequately account for toxic effects which may arise from discharge of complex effluents. Toxicity based consents are being developed for this reason. The application of direct toxicity assessment may be necessary as a control measure for many such effluents. However, for others, once the toxic fraction has been identified, chemical standards may again prove a cost effective control option.

5 CONCLUSIONS

1. EQSs establish concentrations of individual substances in receiving waters which should not be exceeded if specific uses are not to be compromised.
2. When the use is 'protection of the aquatic ecosystem' the definition of a precise objective is problematic and subject to considerable latitude in interpretation.
3. The relevance and reliability of data are the keys to establishing sound standards.
4. Two approaches, statistical models and empirical safety factors are used to extrapolate laboratory effects data to the environment. Both have advantages and disadvantages and require further development. However, transparency of the approach used and an understanding of assumptions made are important.
5. The use of EQSs is likely to continue as an important risk management tool but, in the future, these are likely to be more integrated with ecological monitoring and direct ecotoxicology assessment and are likely to be applied across different environmental media.

References

1. CEC, 'Council Directive on pollution caused by certain dangerous substances discharged into the aquatic environment of the Community (76/464/EEC)' *Official Journal of the European Community* 1976, L129, 18 May.
2. CSTE/EEC, *Rev. Environ. Contam. Toxicol.*, 1994, **137**, 83.
3. OECD, 'Draft Guidance document for Aquatic Effects Assessment'. Prepared for the OECD Hazard Assessment Advisory Body, Organisation for Economic Cooperation and Development (OECD) Paris, February 1992.
4. M. A. Van der Gaag, P. B. M. Stortelder, L. A. Van der Kooy and W. A. Bruggeman, *Europ. Water Pollut. Control*, 1991, **1**, 13.
5. D. Mackay, *Environ. Sci. Technol.*, 1992 **16**, 274.
6. T. Aldenberg and W. Slob, 'Confidence limits for hazardous concentrations based on logistically distributed NOEC toxicity data', RIVM Report No. 71902002, 1991. RIVM, Bilthoven, The Netherlands.
7. E. P. Smith and J. Cairns, *Ecotoxicology*, 1993, **2**, 203.
8. W. Sloof, J. H. Canton and J. L. M. Hermens, 1983, *Aquatic Toxicology* **4**, 113.
9. US EPA, 'Estimating concern levels for concentrations of chemical substances in the Environment'. Environmental Effects Branch, US Environmental Protection Agency, 1984.
10. ECETOC, 'Aquatic Toxicity data evaluation. European Centre for Ecotoxicology and Toxicology of Chemicals', 1993, Technical Report No. 56.
11. OECD, 'Report of the OECD Workshop on the extrapolation of laboratory aquatic toxicity data to the real environment'. 1992, OECD Monograph No. 60, Organisation for Economic Cooperation and Development, Paris 43 pp.

When the Solution is More Toxic than the Pollution

J. H. Churchley and E. M. Hayes

Severn Trent Water Ltd, St Martins Road, Coventry, CV3 6PR, UK

ABSTRACT

The pollution of certain watercourses, especially in the East Midlands, by colour derived from dye residues has led to significant public complaint. In response, the National Rivers Authority (now the Environment Agency) has set tight absorbance consent conditions on nine dye-affected sewage treatment works within the Severn Trent region.

Numerous methods for removal of colour have been proposed and examined at laboratory and pilot scale. In the majority of cases, dye-waste is discharged to foul sewer, and the possibility exists to treat for colour removal on the sewage treatment works. Ozone has proved to be a very effective and versatile treatment, but is relatively high in capital and operating costs.

Chemical treatment using organic polymers to coagulate/flocculate the residual dye can achieve high levels of colour removal. Used in conjunction with existing settlement tanks on a sewage treatment works they offer an attractive, low capital cost, easily retro-fitted colour removal method.

Following extensive jar test programmes, and in response to pressure from the National Rivers Authority to ameliorate the colour problem, Severn Trent Water Limited have embarked on a number of trials with polymers at affected sewage works. In two of these trials, the onset of polymer dosing was accompanied by a rapid rise in final effluent ammonia concentrations to levels above the discharge consent condition.

A number of polymers offered for colour removal by major suppliers were examined for nitrification inhibition using the standard DoE test. A wide range of toxicities was found. Two of the most successful products for removal of colour were found to cause 100% and 70% inhibition at the concentration required for colour removal.

Subsequent examination indicated that despite adequate floc formation and settlement before the biological nitrification process, the toxicity was little reduced. This led to the supposition that water soluble components were responsible for the observed inhibition.

1 INTRODUCTION

Certain rivers in textile producing areas of the UK such as in the East Midlands, have given rise to significant public complaint due to the level of coloration. In 1992 this led to approximately 300 complaints of colour to the National Rivers Authority (NRA) (now the Environment Agency), Severn Trent Region.[1]

Unnatural colour in rivers caused by dye residues is an environmental problem with a complex series of causes. Since the majority of the UK dyers discharge untreated effluent to foul sewer for treatment in admixture with domestic sewage, it is frequently a problem for the Water Company operating the sewage treatment works.

Despite the level of public complaint, the presence of colour is not regarded as toxic or harmful to the watercourse - the pollution is aesthetic only.

Many methods have been examined for colour removal both at the dyehouse and at the receiving sewage treatment works. However, most require considerable capital expenditure which is difficult to provide with a suitable pay-back period.

The use of synthetic organic polymers especially of cationic products has offered an attractive solution to colour removal. Dosing of these materials into the dye-waste or dye-rich sewage offers the potential advantages of excellent colour removal and low sludge production.

This paper describes the use of cationic polymers in trials on sewage treatment works and highlights severe problems with inhibition to nitrification. Further laboratory work has confirmed the high toxicity of some products and has prevented their use in colour removal.

2 BACKGROUND TO COLOUR POLLUTION

Many rivers have shown a major improvement in water quality over the last 20 years. This has led to a much lower water turbidity as well as reductions in BOD and ammonia.[2] Rivers once avoided by the public are now seen as important for leisure activities as well as precious environmental assets. Colour is a highly visible (as well as a highly traceable) pollutant especially in low turbidity waters. No wonder then that an environmentally conscious public with increased leisure time should complain.

The colour pollution of rivers is, however, much more than a problem of increased public awareness and improved river quality. The nature of the dyeing processes, the attitudes of clothing retailers and the fashion-conscious public all play a major role.

The fashion requirement for cotton sportswear and leisurewear in bright and dark colours and the consumers requirement for exceptional wash-fastness has

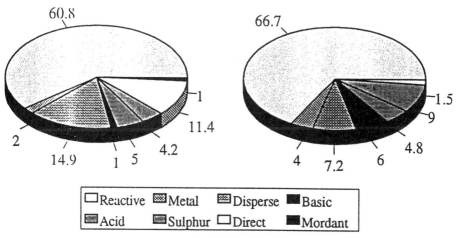

Figure 1 Textile Dyes in Sewage Works Catchments

led to the increased use of reactive dyes. Holme[3] indicates an increased use of reactive dyes for cellulosic fibres from 23,000 tonnes in 1973 to 60,000 tonnes in 1988. Textile producing areas such as the East Midlands in which knitted fabrics predominate may give rise to a high proportion of reactive dyes in the effluent. Figure 1 shows the proportions of dyes of each classification used in the Loughborough and Leicester catchments. Over 60% of all dyes used are reactive dyes in these two catchments.

Reactive dyes pose environmental problems because of their poor fixation in the dyeing process and due to the poor removal on conventional sewage treatment plants. Poor fixation is a common characteristic with reactive dyeing, and although high-fixation dyes are available, they are more expensive and cover only a limited colour range. Estimated degrees of fixation and the percentage loss to effluent are given by Easton[4] and are reproduced in Table 1. Clearly, with dye losses to effluent in the range 10-50%, large quantities of reactive dye find their way into the sewage at certain sewage treatment works.

Table 1 *Estimated Fixation and Loss to Effluent for Different Dyes*

Dye class	Degree of fixation (%)	Loss to effluent (%)
Acid	80–95	5–20
Basic	95–100	0–5
Direct	70–95	5–30
Disperse	90–100	0–10
Metal-complex	90–98	2–10
Reactive	50–90	10–50
Sulphur	60–90	10–40
Vat	80–95	5–20

Table 2 Fish Toxicity Data

LC50 (mg l^{-1})	% of commercial dyes
<1	2
1–10	1
10–100	27
100–500	31
>500	28

Deeper shades result in higher losses of dye to sewer, and Pierce[5] has calculated (and measured) high concentrations of dye (600 mg l^{-1}) in the combined dyehouse effluent from full black dyeing with reactive dyes.

3 ENVIRONMENTAL EFFECTS OF DYES

Clarke and Anliker[6] have reviewed the toxicities of a wide range of dyestuffs. Table 2 shows the results of a survey of fish toxicity data for over 3000 commercial dyes.

Of the dyes showing an LC50 of less than 1.0 mg l^{-1}, the majority were basic dyes. These are not widely used, show excellent fixation on the fibre during the dyeing process, and are removed well by adsorption during conventional sewage treatment.

No chronic toxicity tests on fish have been reported, though bioaccumulation can be expected to be low due to the high water solubility of most dyes reaching the aquatic environment. Even low water solubility disperse dyes have been shown not to bioaccumulate.[7]

Brown[8] has predicted likely concentrations of dyes in rivers. Average estimates for daily dye usage, fixation, removal in sewage treatment and dilution in the receiving water suggest a concentration of 1.0 mg l^{-1} for a single dye in a typical river. Hobbs[9] calculations show that for a 'worst case' scenario the likely river concentration of reactive dye would reach 1.5 mg l^{-1}.

At concentrations likely to occur, the environmental effect of the vast majority of dyes is thus one of aesthetic pollution rather than of toxicity to aquatic life.

4 COLOUR REMOVAL

Modern dyes, developed for their stability and fastness exhibit low degrees of biodegradation in the aerobic environment of biological sewage treatment.[10,11] However, some dye types are removed in conventional sewage treatment by processes such as coagulation/sedimentation and adsorption onto sludge solids. Hitz et al[12] applied a laboratory test to examine the effectiveness of adsorption on activated sludge as a mechanism for removal of dyestuffs in six classes. The results may be summarised:

Acid dyes — low adsorption
Basic dyes — high adsorption
Direct dyes — high adsorption
Disperse dyes — high to medium adsorption
Reactive dyes — very little adsorption.

Significantly, very little adsorption of reactive dyes takes place on activated sludge; on percolating filters with less prolonged contact between biomass and substrate, even less removal is likely.

In areas where reactive dyes predominate, therefore, little colour removal takes place on conventional sewage treatment works. The East Midlands region is further characterised by relatively low river flows,[9] giving low dilution to coloured effluents.

The NRA (now the Environment Agency) has responded to the coloration of rivers and to the level of public complaint by setting colour consent conditions for dischargers of potentially coloured wastewater into the affected rivers. These consent conditions are based on absorbance measurements on pre-filtered samples (0.45 micron) at a number of wavelengths (varying from 3 to 23) across the visible spectrum. The setting of the colour standards is described by Waters.[13] Severn Trent Water Ltd currently have ten such colour standards applied to final effluents from various sewage works.

Colour standards require the application of specific colour removal measures to be adopted at the dye-house or at the sewage works.

In general, colour removal treatment strategies are based on the general methods given in Table 3. Many of these methods are relatively unproven. However, the use of organic polymers has been proven to be effective in plants treating dye-waste for direct discharge to river.[13, 14] These plants utilise dissolved air flotation to separate colour flocculated by cationic polymer following biological treatment. A major advantage of organic polymers for coagulation/flocculation treatment is the low quantity of sludge produced. These polymers, if used on a sewage treatment plant to flocculate dye-waste colour with settlement of the insolubilised colour in existing sedimentation tanks, would give an easily retro-fitted colour removal process with very low capital cost.

5 TRIAL AT WORKS A

Works A is situated in Nottinghamshire close to the Derbyshire border and discharges treated effluent to the River Erewash, a tributary of the Trent. Flow to the works is approximately 4,000 m^3 d^{-1} in dry weather of which some 2,000 m^3 d^{-1} is derived from a fabric dye-house. The dye-waste is received at the STW in a dedicated sewer and is pre-treated on a Vitox oxygen activated sludge plant before conventional treatment in admixture with domestic sewage. A schematic diagram of the flowsheet is shown in Figure 2.

Table 3 *Colour Removal Treatment Methods*

Treatment Type		Comments
1. Oxidation	1.1 Ozone	Very effective. High capital and revenue cost.[14]
	1.2 Chlorine dioxide	Relatively unproven. Possible toxic effects.
2. Reduction	2.1 Anaerobic biodegradation	Only for azo dyes. Possible toxic by-products and possible odour.
	2.2 Chemical reduction	Possible toxic by-products. Uncertain performance.
3. Coagulation/ flocculation	3.1 Inorganics e.g Fe^{II}, Fe^{III}, Mg^{II} salts	Moderately effective, high sludge production. Used by a few UK dyers.[15]
	3.2 Organic polymer	Moderately effective, low sludge production. Used by a number of UK dyers.[16,17]
4. Adsorption	4.1 Inorganic adsorbent	Good performance as part of water re-use
	4.2 Organic adsorbent	Relatively unproven. May form part of water re-use scheme.
5. Other methods	Ultra-filtration electrolysis etc.	Ultra-filtration may form part of water re-use scheme.[18]

The Works has attracted a colour consent from the NRA (now Environment Agency) which has been in force since 1 January 1996. Earlier notification of this colour standard caused Severn Trent Water to investigate various methods of colour removal appropriate to the situation. Principal among these was the use of cationic organic polymers. The major polymer suppliers were approached and all visited the plant to carry out jar tests. One material, Polymer 1, performed extremely well in jar tests, giving adequate colour removal to meet the colour standards.

A trial was arranged with Polymer 1 dosed directly into the Vitox return activated sludge (point X on Figure 2) at a dose of 100 mg l^{-1} calculated on dye-waste flow. The trial was started in October 1993, prior to the dyers busy pre-Christmas period. From the start, colour removal was not as good as had been obtained in jar tests. A change in the nature of the dye-waste or poor mixing at the dosing point were thought to be to blame for the poor colour removal. After two weeks, no improvement was found, and the dose was increased to 200 mg l^{-1}, a dose which had given even better performance than 100 mg l^{-1} in jar tests.

After a few days at the higher dose rate, high final effluent ammonia concentrations were observed. The polymer dose was stopped and the works

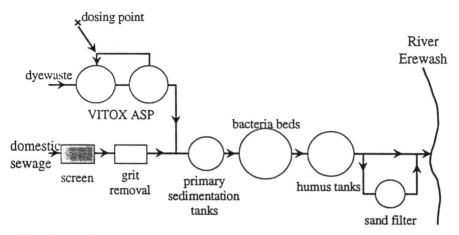

Figure 2 *Schematic Diagram – Works 'A'*

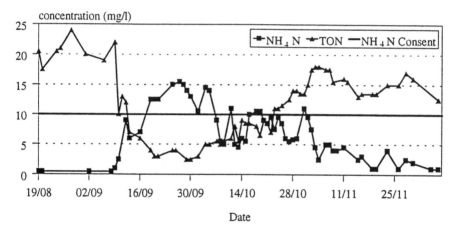

Chemical dosing trials started 09/09 100ppm, 20/09 - dose increased to 200ppm, trial ceased 25/09.

Figure 3 *Works 'A' Final Effluent*

nitrification performance closely observed. Data for this period of polymer dosing and for the recovery period are plotted in Figure 3.

The ammonia concentration rose very rapidly at the start of polymer dosing. The corresponding fall in total oxidised nitrogen (TON) shows that the nitrification process was inhibited rather than a high influent ammonia being the cause. The ammonia concentration breached the 10 mg l^{-1} ammonia nitrogen standard resulting in consent failures. Recovery was very slow despite re-seeding the bacteria beds with nitrifying activated sludge.

Subsequent consideration of the events led to the conclusion that poor flocculation conditions (which contributed to poor colour removal) gave rise to un-flocculated polymer carried over onto the bacteria beds where the nitrification process was inhibited. The apparently high toxicity was surprising since Polymer 1 had been used on a number of biological treatment plants with no problems reported.

6 LABORATORY TEST WORK

Following the trial, a series of laboratory tests were carried out to examine the acute nitrification inhibition by a number of organic polymers offered for colour removal. The test method used was the Department of the Environment (DoE) Standard method.[19] This method requires the use of a good quality nitrifying activated sludge which is divided into a number of aliquots. These are aerated with an ammonia source with and without the test substance. Ammonia and nitrate are measured before and after four hours aeration. Nitrification inhibition is determined by comparing the performance of the blank and test aliquots. Details of the experimental procedure are given below.

6.1 Experimental Procedure

A sample of nitrifying activated sludge was taken from a fully nitrifying activated sludge plant and was maintained in an aerobic condition prior to use. The sample of activated sludge was allowed to settle and the supernatant was discarded. The activated sludge solids were washed twice with distilled water and re-suspended in distilled water to give a suspended solids of approximately 3000 mg l^{-1}. The washed activated sludge was divided between six test vessels, with 250 ml in each.

Settled sewage, sampled from the same works, was divided into six aliquots of 250 ml each. Dilutions of the test polymer were made into the 250 ml settled sewage aliquots including two blanks with no test polymer.

Settled sewage and activated sludge were mixed and samples were taken for initial ammonia and TON analysis. Temperature and pH were measured. Temperature was room temperature and showed little change during the test. Initial pH was always in the required range of seven to eight. The mixtures were aerated for four hours using aeration stones and aquarium air pumps - this aeration ensured that dissolved oxygen remained above 2.0 mg l^{-1} throughout the test, and that all suspended solids remained in suspension.

After four hours aeration, the flask contents were filtered through a GFC paper and the filtrates submitted to the laboratory for ammonia and TON analyses. The degree of inhibition was calculated for each polymer test concentration:

$$\text{Degree of inhibition (\%)} = \frac{(C_c - C_i)}{C_c} \times 100$$

where C_c = concentration TON formed in control
C_i = concentration TON formed in presence of test substance.

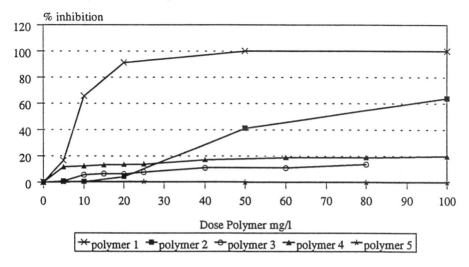

Figure 4 *Nitrification Inhibition by Polymer. Results of Laboratory NI Test*

6.2 Experimental Results

Results of nitrification inhibition tests are shown in Figure 4. Clearly, the results of the laboratory test support the findings of the plant trial at Works A - Polymer 1 is completely inhibitory at concentrations of 50 mg l^{-1} and above. Polymer 5 is non-toxic to the nitrification process at all concentrations tested. All the other materials exhibit a toxicity intermediate between Polymer 1 and Polymer 5. Significant among these other materials is Polymer 2 showing 64% inhibition at 100 mg l^{-1}.

7 CONCLUSIONS

Colour pollution of certain watercourses is unsatisfactory and measures to curb unnatural coloration are required. However the pollution is aesthetic only and no toxic effects to aquatic life have been reported at likely dye concentrations in the river.

One of the colour removal methods most likely to be viable at the sewage treatment works is chemical coagulation/flocculation using organic synthetic polymers. The use of existing settlement tanks makes the process low in capital cost, and most likely to be financially acceptable to the textile dyers. Sludge production is low and no pH correction is needed.

However, results presented show that some polymers are highly toxic to nitrification processes at concentrations similar to those used for colour removal. Before these products are used, extensive nitrification inhibition tests should be carried out to avoid ammonia pollution by the inhibition of nitrification.

Cationic polymers are highly toxic to fish with acute static LC50 values ranging from 0.090 to 0.40 mg l^{-1} for freshwater species.[20] With analytical techniques unable to detect concentrations at LC50 levels, the addition of cationic polymers to prevent an aesthetic pollution is difficult to justify environmentally.

ACKNOWLEDGEMENTS

The valuable contribution of Anna Collins to the trial and laboratory test work is gratefully acknowledged. Continued support and co-operation by Mark Betts and Geoff Gent at Pinxton is much appreciated. Views expressed are those of the authors and not necessarily those of Severn Trent Water Ltd..

References

1. G. Morris and B. D. Waters, The impact of textile trade effluents on water quality, IWEM symposium 'Textile Industry Trade Effluents', February 23rd, 1993, Rochdale, Lancs.
2. J. R. Martin, and D. J. Brewin, Quality improvements and objectives in the Trent catchment, ICE/IWEM conference 'Wastewater treatment - what's it worth', March 9-10th, 1994, London.
3. I. Holme, *Rev. Prog. Coloration*, 1992, **22**, 1.
4. J. R. Easton, 'Colour in Dye-house Effluent', Ed. P. Cooper, Society of Dyers and Colourists, Bradford, 1995, Chapter 1, p. 9.
5. J. Pierce, *J. Soc. Dyers & Colourists*, 1994, **110**, 131.
6. E. A. Clarke and R. Anliker, *Rev. Prog. Coloration*, 1984, **14**, p. 84.
7. R. Anliker and P. Moser, *Ecotox. Environ. Safety*, 1987, **13**, p. 43.
8. D. Brown, *Ecotox. Environ. Safety*, 1987, **13**, p. 139.
9. S. J. Hobbs, *J. Soc. Dyers & Colourists*, 1989, **105**, p. 355.
10. U. Meyer, Biodegradation of synthetic organic colorants, FEMS symposium, 1981, Vol. 12, p. 371.
11. U. Pagga and D. Brown, *Chemosphere*, 1986, **15**, no. 4, p. 479.
12. H. R. Hitz, W. Huber and R. H. Reed, *J. Soc. Dyers & Colourists*, 1978, **94**, p. 71.
13. B. D. Waters, 'Colour in Dye-house Effluent', Ed. P. Cooper, Society of Dyers and Colourists, Bradford, 1995, Chapter 2, p. 22.
14. J. H. Churchley, *Wat. Sci & Tech.*, 1994, **30**, no. 3, p. 275.
15. E. J. Newton, 'Colour in Dye-house Effluent', Ed. P. Cooper, Society of Dyers and Colourists, Bradford, 1995, Chapter 8, p. 101.
16. J. Scotney, 'Colour in Dye-house Effluent', Ed. P. Cooper, Society of Dyers and Colourists, Bradford, 1995, Chapter 7, p. 92.
17. Anon, *Ind. Waste Management*, 1992, November, p. 22.
18. C. A. Buckley, *Wat. Sci. Tech.*, 1992, **25**, No. 10, p. 203.
19. Department of Environment, The assessment of the nitrifying ability of activated sludge (Tentative methods), 1980, HMSO, London.
20. C. H. Wolf, 'Toxicology of cationic water soluble polymers', MSc thesis University of San Francisco, Environmental & Occupational Toxicology, September, 1983

Possible Ways of Detoxification of Heavy Metals (Zinc, Lead and Cadmium) Discharged with Wastes in High-Bioproductive Reservoirs

I. V. Iskra and P. N. Linnik

Department of Hydrochemisty, Institute of Hydrobiology, Geroev Stalingrada Ave. 12, Kyiv, 254210, Ukraine

ABSTRACT

This paper considers the problem of elevated levels of some heavy metals in the Dnieper reservoir system. The complexing ability of dissolved organic matter to detoxify heavy metals is investigated. The authors conclude that, for some metals, the complexing of heavy metals with dissolved organic matter in high-bioproductive systems is the dominant factor of detoxification.

1 INTRODUCTION

Heavy metal pollution is a problem of great importance for Ukrainian surface water, especially for the Dnieper reservoirs which play a significant role in the Ukrainian national economy. The cascade of six reservoirs with a total volume of 44 km^3 provides the drinking water for up to 75% of the population of the Ukraine and irrigate more than 1 million hectares of agriculture land. They are a source of cheap hydro-electric energy and form a major transportation artery.

The main source of heavy metals, especially in the most polluted southern reservoirs, is insufficiently treated wastes from mining, metallurgic and chemical factories.

Toxicity of the water environment for water organisms and finally for man depends primarily not on total content but the physico-chemical state of heavy metals in water.[1-3] Free (hydrated) ions are regarded as the most toxic forms.[2,6] The high toxicity of industrial wastes containing heavy metals is caused basically by free ions. The binding of heavy metals in complexes with inorganic, and particularly with naturally occurring organic matter, more often than not causes a decrease in toxicity or complete suppression.[2,7]

A peculiarity of the Dnieper reservoirs is their high bioproductivity and the wide diversity of composition of the dissolved organic matter (DOM). The content of C_{org} is on average 6.7-12.7 mg l^{-1}.[8] DOM mainly consists of humic compounds.

2 MATERIAL AND METHODS

The water samples were taken during boat expeditions along the Dnieper reservoirs in different seasons. Immediately after sampling, they were filtered using membrane filters ("Synpor" Czech republic) with pore diameter 0.4 μm. Total concentration of the dissolved forms was determined after UV irradiation in quartz glass. Suspended forms of heavy metals were measured after wet combustion in a mixture of nitric and sulphuric acids.

For determination of the complexing ability of DOM, in aliquots of volume 0.5-1.0 litre, different concentrations (50–1000 μg l^{-1}) of zinc, lead and cadmium were added. The concentrations were controlled just after metal addition, during some hours and then once a day till reaching equilibrium in the system. The concentration of unbound metal complexes (free ions) was measured using high-sensitive Anodic Stripping Voltammetry.[4] This method permits the detection of the most toxic form of heavy metals, their free ions,[5] using a polarographic analyser (PA-2 Czech republic). Possible concentration changes owing to adsorption on vessel walls, hydrolysis and subsequent coprecipitation were determined in control experiments using doubly distilled water adjusted to natural level pH.

Molecular mass and chemical nature of the complex compounds of heavy metals were determined using gel-permeation chromatography on neutral sephadexes (G-75, Pharmacia, Sweden; HW-50, Toyopearl, Japan) and ion exchange chromatography.

3 RESULTS AND DISCUSSION

The water ecosystem has to cope with the continuously increasing levels of heavy metals. In most cases, reducing the impact of heavy metal pollution is a result of transformation of toxic forms to less active forms. The degree of detoxification depends first of all on the intensity of intrawater body processes, which favour reducing the concentration of free heavy metal ions as the most toxic form. Among such processes are:

1. adsorption on suspended particles;
2. complexing (mainly with ligands of organic origin);
3. sedimentation and coprecipitation;
4. hydrolysis and forming of low soluble compounds;
5. accumulation and adsorption by biota.

Table 1 *Suspended Forms of Zinc, Lead and Cadmium in the Kremenchug and Kakhovka Reservoirs in Different Seasons (1990-1994), % of Total Content*

Seasons	Kremenchug reservoir			Kakhovka reservoir		
	Zn	Pb	Cd	Zn	Pb	Cd
Spring	28.6	58.4	22.0	32.3	60.3	15.0
Summer	38.5	80.8	15.1	36.6	66.4	16.0
Autumn	23.4	54.5	34.0	28.8	55.4	–

It is difficult to prioritise the above-mentioned processes because in different water bodies (by type and hydrological regime) each can play a dominant role in the transformation of heavy metal compounds.

3.1 Adsorption on Suspended Particles

Intensity of adsorption and its contribution to detoxification in the water environment depends on the water regime, turbidity and origin of suspension. Moreover, the chemical properties of individual heavy metals is of great importance. For example, at the same hydrological conditions such metals as Cd, Cr, Cu, Zn mainly form dissolved compounds, others, Pb, Mn, more easily migrate in suspended state.[2] In the high-bioproductive Dnieper reservoirs this tendency is confirmed. The dominant form of existence for Zn and Cd is their dissolved compounds (Table 1). This is conditioned by low turbidity of the water and high content of DOM. The maximum turbidity of Dnieper water is observed in Spring-Summer time (5-25 mg l^{-1}); its minimal values in Autumn (3-10 mg l^{-1}) and Winter (1-2 mg l^{-1}).[8] The seasonal variations of turbidity show a good correlation with the content of the suspended form. The quantity of suspended metals in the water of the reservoirs more often than not rises in the Summer period due to the intensive growth of phytoplankton.

Thus, adsorption on suspended particles can play a definite role in detoxification in the aquatic environment. It is especially essential in the case of Pb, which absorbs relatively easily onto suspended matter of organic and inorganic origin. For Zn and Cd, adsorption is less.

3.2 Complexing as a Factor of Detoxification

Dissolved forms of Zn, Pb and Cd in water of the Dnieper reservoirs are mainly present in the form of complexes with organic matter. Depending on the season, the degree of heavy metal complexation varies, reaching a maximum of 66-98% of the total content of dissolved forms (Table 2). The minimum values were observed in Spring, when the phytoplankton begins to grow and the reservoir water is considerably diluted by snow melt water. As our investigation showed, complexing in the Dnieper reservoirs is a highly intensive process with participation of naturally occurring DOM, both of

Table 2 *Binding Degree of Zinc, Lead and Cadmium in Complexes with DOM in the Dnieper Reservoirs (1990-1994), % Concentration of Dissolved Metal*

Reservoirs	Spring			Summer			Autumn		
	Zn	Pb	Cd	Zn	Pb	Cd	Zn	Pb	Cd
Kyiv	72.8	88.5	77.0	86.4	98.3	82.5	90.6	97.6	85.2
Kremenchug	68.5	90.7	73.0	81.7	97.9	78.1	86.8	95.0	87.8
Kakhovka	66.0	84.6	72.2	84.9	96.8	74.7	88.0	93.4	84.8

Table 3 *Complexing Ability (CA)* of Dissolved Organic Matter of the Kyiv, the Kanev and the Kremenchug Reservoirs, $\mu g\ l^{-1}$*

Metal by which CA is determined	Kyiv reservoir	Kanev reservoir	Kremenchug reservoir
Cd (II)	38.1–132.2	31.4–147.8	–
Pb (II)	66.2–240.1	231.8–333.2	256.7–382.9
Zn (II)	74.1–219.0	109.2–124.8	101.4–226.2

* – expressed by quantity of metal bound in complexes

allochthonous and autochthonous origin. Complexing plays the dominant role in detoxification of heavy metal compounds in the Dnieper reservoirs. The investigation was aimed to determine the complexing ability of DOM.

3.2.1 Complexing Ability of DOM Complexing ability allows an evaluation of the maximum possible total concentration of organic ligands and, more precisely, their free co-ordination centres which can bind heavy metal ions in complexes. The quantitative expression of complexing ability is a concentration (in $\mu mol\ l^{-1}$ or $\mu g\ l^{-1}$) of the part of metal which pass to the bound state in condition of "saturation" of active complexing centres by metal ions. Complexing ability should be determined by the concentration of different metals, taking into account their chemical properties and their ability to form complexes.

Data about the complexing ability of DOM in the Dnieper reservoirs are shown in Table 3. Cadmium binds in complexes in the lowest degree. It means that after discharging the wastes containing heavy metals, the toxic forms of cadmium will exist in the water environment for longer than compounds of zinc or lead. Complexing ability undergoes seasonal changes (Figure 1), that is conditioned by the different component composition of DOM in the Dnieper reservoirs. Minimal quantities of metals are bound in complexes in Spring. Most likely this is due to prevailing high molecular humic compounds of allochthonous origin in that period. As it is known,[9] high molecular organic matter has negligible complexing properties. In the Summer-Autumn period,

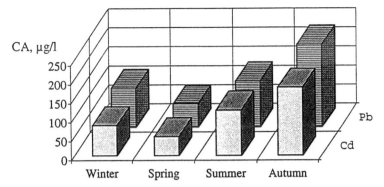

Figure 1 *Complexing Ability of DOM in Different Seasons in Water of the Kyiv Reservoir (Average Values).*

complexing ability is greatest. In this period, the organic compounds, products of metabolism, take part in complexing side by side with humic compounds (mainly fulvic acids).

The data presented suggest that the water ecosystem is more resistant to heavy metal pollution and can more easily detoxicate wastes containing heavy metals in Autumn. In Spring, when values of complexing ability are minimal, the water environment is most vulnerable to heavy metal pollution.

3.2.2 Kinetics of Complexing The reliable determination of complexing ability can be ensured by taking into consideration the complexing kinetics and reaching equilibrium in system. We determined that "saturation" of active co-ordination centres is reached when the metal is added at a concentration of 0.5-1.0 mg l^{-1}. Equilibrium in the system is established over a period of several days. From our point of view, these conditions are optimal because they neglect the destruction of DOM for this short period of time and reliably determine the potential complexing ability.

As the results of our previous investigation showed, velocity of complexing is different for heavy metals. The equilibrium for Mn^{2+} and Cd^{2+} is reached very slowly (8-20 days), while for Cu^{2+} and Cr^{3+} the velocity of complexing is the highest (from some hours to 1-3 days).[10] An example of the kinetics of complexing for Zn and Cd in the Kyiv reservoir is presented in Figure 2.

According to the law of mass action, and our results, the concentration of the added metal can significantly influence the kinetics of complexing. It proceeds very slowly when concentrations of added metal reach 50–100 μg l^{-1}. The high level of added metal is limited by the hydrolysis processes. The coefficient of complexing velocity for reactions of the first order (because the total concentration of organic ligands is constant, and many times higher than the concentration of added metal) at constant temperature is in the form of:

$$k = \frac{1}{\tau}\ln\frac{C_0}{C_\tau}$$

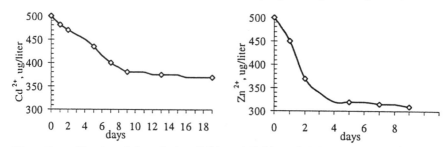

Figure 2 Kinetics of Complexing Cd^{2+} and Zn^{2+} with DOM in Water of the Kyiv Reservoir. Summer 1994; $[Cd^{2+}] = [Zn^{2+}] = 500 \ \mu g \ l^{-1}$

where, k - coefficient of complexing velocity; τ - time after beginning the reaction; C_0 - initial metal concentration; C_τ - metal concentration at time τ.

The next step is to calculate when 90% complexing will occur, or almost full detoxification of the metal:

$$\tau_{90\%} = \frac{2.3}{k}$$

According to our investigation for Zn and Pb the time to achieve 90% binding is 9 - 18 days, for Cd it is 22 - 42 days.

3.2.3 Chemical Nature of Complex Compounds. We made an attempt to investigate the role of different types of organic matter in binding heavy metal ions in complexes. After equilibrium was reached, the water was fractionated on columns with ion-exchange cellulose. Three fractions of DOM were obtained - acid, base and neutral. The first includes mainly humic compounds (humic and fulvic acids), phenols and carboxylic acids. The second fraction is made up mainly by protein-like compounds (proteins, polypeptides). The third group is represented by polysaccharides and free reducing saccharides.

The data on Cd complexes and their chemical nature in Summer time are shown in Figure 3. In original water, the major part of the investigated metal (58-82% $Cd_{complex}$) is in the fraction of humic compounds (anion fraction of DOM). Rather less complex compounds of Cd are in the fraction of carbohydrates (neutral group) and protein-like compounds (cation group). After reaching the equilibrium of complexing, the portion of Cd with humic substances becomes less in both seasons. The data we obtained for Zn and Pb show the same tendency: in conditions of "saturation", humic compounds do not increase actively bound heavy metals, the organic matter of cation and neutral origin joins the processes of complexing.

3.2.4 Molecular-mass Distribution of Complex Compounds We conducted a special investigation to study molecular mass distribution of heavy metal complex compounds in original water, and after addition of heavy metals, reaching complexing equilibrium. The main aim of this investigation was to

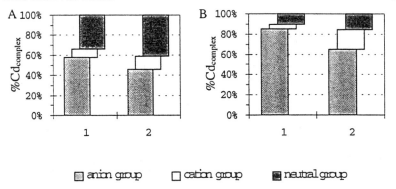

Figure 3 *Chemical Nature of Complex Compounds of Cd in Original Water (1), and After Reaching Complexing Equilibrium (2). A-Winter, B-Autumn. Kyiv Reservoir*

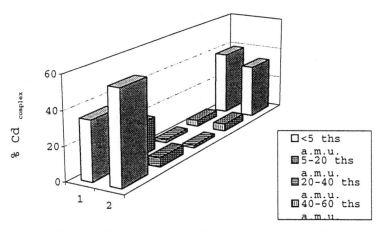

Figure 4 *Distribution of Cd Among Complex Compounds of Different Molecular Mass in Original Water (1), and After Reaching Complexing Equilibrium (2). Winter, 1995. Kyiv Reservoir. Complexing Ability = 65 µg l^{-1}*

determine the molecular-mass of the type of organic matter which was playing the most active rôle in complexing. An example of Cd (Figure 4) showed that in condition of "saturation" of active co-ordination centres by heavy metals ions, the complexing proceeds mainly at the expense of relatively low molecular mass ligands (< 5 ths a.m.u.). Approximately the same distribution was obtained for Zn and Pb complex compounds.

4 CONCLUSION

Detoxification of the investigated heavy metals discharged with wastes occurs by the transformation of toxic forms to less active forms and is determined by the whole complex of intrareservoir processes. In the high-bioproductive reservoirs, such as the Dnieper reservoirs, complexing with DOM of natural origin is the dominant factor of detoxification. High values of complexing ability of DOM demonstrate high potential of the Dnieper water to bind heavy metals in complexes and, as a result, to reduce heavy metal toxicity. However in Spring time complexing ability is lowest and this can cause worsening of the water quality owing to increasing concentrations of free metal ions, the most toxic form of heavy metals.

References

1. M. Florence and G. E. Batley, *CRC Crit. Rev. Anal. Chem.*, 1980, 9, 219.
2. N. Linnik and B. I. Nabivanets, 'Migration forms of metal in surface water', Gidrometeoizdat, Leningrad, 1986 (in Russian).
3. G. Pagenkopf, 'Type of metal ions and its toxicity on water systems / Concepts on metal ion toxicity', Mir, Moscow, 1993 (in Russian).
4. P. N. Linnik and I. V. Iskra, *Microchem. J.*, 1994, 50, 184.
5. G. M. Morrison and Ch. Wei, *Anal. Proc.*, 1991, 28, 70.
6. S. N. Luoma, *Sci. Total Environ.*, 1983, 28, 1.
7. J. W. Moore and S. Ramamoorty, *'Heavy metals in natural water'*, Mir, Moscow, 1987 (in Russian).
8. A. I. Denisova et al., *'Hydrology and hydrochemistry of the Dnieper and its reservoirs'*, Naukova Dumka, Kyiv, 1990 (in Russian).
9. F. H. Frimmel, A. Immerz and H. Niedermann, *Intern. J. Env. Anal. Chem*, 1983, 14, 105.
10. P. N. Linnik et al., *Hydrobiol. J.*, 1994, 30, N5, 87 (in Russian).

United States Environmental Protection Agency's Water-quality Based Approach to Toxics Control

Margarete A. Heber[1] and Teresa J. Norberg-King[2]
Editorial synopsis: John Tapp[3], James Wharfe[4] and
Stephen Hunt[5]

US Environmental Protection Agency, Office of Science and Technology, Washington, DC, USA
[2]US Environmental Protection Agency, Mid-Continent Ecology Division, Duluth, MN, USA
[3]Brixham Environmental Laboratory, Zeneca Ltd, Freshwater Quarry, Brixham, Devon TQ5 8BA, UK.
[4]Environment Agency, Guildbourne House, Chatsworth Rd, Worthing, W. Sussex BN11 1LD, UK.
[5]Hyder Environmental, Howard Court, Manor Park, Runcorn Cheshire, WA7 1SJ, UK.

Margarete Heber presented an overview of the Water Quality Based Approach to Toxics Control adopted in the United States. Due to other commitments, the authors were unable to provide a paper for the conference proceedings but agreed to the inclusion of the detailed overheads with introductory commentary and explanatory text from the authors and the editors.

This contribution from the United States is likely to be of considerable interest to the readers of these proceedings.

1 TOXICS CONTROL IN THE UNITED STATES

For the protection of aquatic life, the United States Environmental Protection Agency (US EPA) employ an integrated approach for the control of toxic inputs. This approach involves chemical, toxicological and biological assessment. Used in isolation, each component can provide useful but incomplete information.

The purpose of this paper is to give an overview of how the US EPA's water-quality based program works; including how whole effluent toxicity

(WET) and toxicity reduction evaluations (TREs) fit into the water program; followed by more detailed information on WET and TREs.

Two pieces of federal legislation govern most water law in the US; the Safe Drinking Water Act and the Clean Water Act. The Clean Water Act (CWA) is the legal authority under which WET testing falls. From the CWA, and it's amendments, there are specific regulations which more clearly lay out the requirements of the Act. For example, point source discharges are specifically regulated under the CWA via the National Pollutant Discharge Elimination System (NPDES) permits, as are Effluent Guidelines and State Water Quality Standards, as well as Enforcement.[1-15]

Slide 1

Control of Point Source Pollution

- Effluent Guidelines
- Water Quality Standards
- National Pollutant Discharge Elimination System (NPDES)
- Enforcement

Under the CWA, all facilities discharging into the waters of the US must have a NPDES permit. Permits are written based on Effluent Guidelines, State Water Quality Standards, and WET.

Slide 2

Effluent Guidelines

- Technology basis for point source controls
- For different types of industries,
  e.g. paper, petroleum, metal finishing
- Best Available/ Practicable Technology (BAT, BPT)

Effluent Guidelines are the technology basis for point source controls, and exist for many different types of industries (e.g. pulp and paper, petroleum, and metal finishing). Permit limits written on effluent guidelines are based on the Best Available Technology (BAT) or Best Practicable Technology (BPT).

Slide 3

Federal Water Quality Criteria and Standards

- State Standards
- Criteria
- Designated use of the Water Body
- Anti-degradation statement

Water Quality Standards are required under the CWA, but adopted by the individual states as state law. State Water Quality Standards are comprised of a narrative statement (no toxics in toxic amounts), numeric Water Quality Criteria, the designated use of a water body (e.g. drinking water source, fishery), and an anti-degradation statement. State Water Quality Standards are the water quality basis for NPDES permits, beyond the technology basis of effluent guidelines. In addition, WET is part of state Water Quality Standards via the narrative statement, and sometimes as a numeric criterion.

Slide 4

Enforcement

- Violation of NPDES Permit Limit
- Penalties and Fines
- Consent Decree
- Toxicity Reduction Evaluation (TRE) / Toxicity Identification Evaluation (TIE)

Under the authority of the CWA, any violation of an NPDES permit requires an enforcement action of some type. Enforcement action range from a phone call or a letter, to penalties and fines and/or consent decrees. TREs and Toxicity Identification Evaluations (TIEs) may also be part of the enforcement process if a WET permit violation occurs.

Slide 5

EPA's Integrated Water Quality-Based Program

- For the protection of aquatic life:
  chemical specific controls
  whole effluent toxicity controls
  biological criteria/bioassessment
- These complement each other

The use of chemical specific standards alone in permit conditions may fail to provide the necessary protection for aquatic life. The limitations of this approach are identified.

Slide 6

EPA's Strategy for Controlling Effluent Toxicity

Use of biological techniques to complement chemical specific analyses for permit limits.

Why chemical parameters are limiting in regulating toxicity:
- complex chemical analyses

- chemical/biological/physical reactions with other chemicals aren't evaluated.
- chemical data must be evaluated
- inability to predict toxic effects

The use of direct toxicity measures helps overcome some of these limitations

In the United States, toxicity based criteria are embodied in Federal Law. WET is the US EPA's Whole Effluent Toxicity testing programme. It is implemented via the narrative part of the State Standards which embodies the following statement:

"It is the national policy that the discharge of toxic pollutants in toxic amounts be prohibited . . ."

Source: Federal Water Pollution Control Act of 1972; amended by the Clean Water Act of 1987. The mechanism for ensuring this goal is the National Pollutant Discharge Elimination System (NPDES).

WET is an integral part of the US EPA's program. The reasons why it was developed and why it is important were summarised.

Slide 7

What is WET ?
Whole Effluent Toxicity - total toxic effect measured on aquatic organisms for both acute (short) and long-term (chronic) effects.

Why was it developed?
In the early 1980's, effluents were acutely toxic, although permitted discharges were meeting the technology-based chemical limits.

Why is it important?
Difficult to develop chemical specific-criteria for every compound discharged in effluents.

There are a number of advantages in using biological systems to test for environmental impact.

Slide 8

Advantages of Using Biological Testing
- Toxic effects of effluent discharges measured directly
- Bioavailability of toxics assessed
- Interactions of constituents measured
- Complex and costly analytical methods not required

The following are criteria for species selection for permitting decisions.

Slide 9

Criteria for Species Selection

- Availability
- Ease of culture and maintenance
- Background information
- Sensitivity
- Ecologically representative
- Regulatory requirements
- Type of exposure
- Type of effluent sample
- Frequency of sample collection
- Test location

Followed by the considerations for selecting the test protocol.

Slide 10

Considerations in Selecting a Test Procedure

- Species sensitivity
- Species availability
- Dilution water
- Test temperature
- Effluent variability

No one species is always the most sensitive for all chemicals or effluents. At a minimum US EPA requires a fish, an invertebrate and a plant species to be tested.

The acute and chronic methods that have been promulgated into regulation (Oct 16 1995) as the test methods which must be used in permit limits nation wide are listed in Slides 11 and 12. There are additional acute methods (not listed) which are considered "supplemental methods."

Slide 11 *Acute Toxicity Test Methods / Mortality Tests*

Type of Organism	Scientific Name	Test Duration
Freshwater Species		
Invertebrates		
Cladoceran	*Ceriodaphnia dubia*	24, 48; optional: 96 h
	Daphnia magna	24, 48; optional: 96 h
	Daphnia pulex	24, 48; optional: 96 h
Fish		
Fathead minnow	*Pimephales promelas*	24, 48, 96 h

Rainbow trout	*Oncorhynchus mykiss*	24, 48, 96 h
Brook trout	*Salvelinus fontinalis*	24, 48, 96 h

Marine/Estuarine Species

Invertebrate		
Mysid shrimp	*Mysidopsis bahia*	24, 48, 96 h
Fish		
Sheepshead minnow	*Cyprinodon variegatus*	24, 48, 96 h
Silverside	*Menidia beryllina*	24, 48, 96 h
Silverside	*Menidia menidia*	24, 48, 96 h
Silverside	*Menidia peninsulae*	24, 48, 96 h

Slide 12 *Short-term Chronic Toxicity Methods*

Freshwater Species

Species Common name	Test Duration	Test Endpoint(s)
Ceriodaphnia dubia Cladoceran	~7 d	Survival, reproduction (60% have 3 broods)
Pimephales promelas Fathead minnow	7 d	Survival, weight
	7-9 d	Survival, percent hatch, percent abnormal
Selenastrum capricornutum algae	96 h	Growth

Marine/Estuarine Species

Species Common name	Test Duration	Test Endpoint(s)
Arbacia punctulata sea urchin	1.5 h	Fertilisation
Champia parvula	7-9 d	Cystocarp production (fertilisation)
Mysidopsis bahia	7 d	Survival, growth, fecundity
Cyprinodon variegatus Sheepshead minnow	7 d	Survival, weight
	7-9 d	Survival, percent hatch, percent abnormal
Menidia beryllina Inland silverside	7 d	Survival, weight

2 TOXICITY REDUCTION EVALUATION

Violation of the permit condition introduces enforcement action. This can involve penalties and fines and the need for Toxicity Reduction Evaluation and Toxicity Identification Evaluation.

Slide 13

Toxicity Reduction Evaluation (TRE)

A study to determine abatement measures to meet toxicity requirements.

Two Approaches to Toxicity Reduction Evaluations (TREs):
- Toxicity Identification Evaluations (TIEs)
- Effluent treatability

The US EPA has developed methods intended to aid dischargers and their consultants in conducting Toxicity Identification Evaluations (TIEs) as part of Toxicity Reduction Evaluations (TREs). Two EPA TRE manuals (EPA, 1989A; 1989B) cover the TIE Phase 1 Characterisation steps. The TIE uses both the acute toxicity test and the short-term chronic test methods that are shown in slides 11 and 12.

Slide 14

TIE has three major parts:
- Phase I: Characterisation:
- Phase II: Identification:
- Phase III: Confirmation:

The TIE approach is divided into three phases. Phase I contains methods to characterise the physical/chemical nature of the constituents which cause toxicity. Such characteristics as solubility, volatility and filterability are determined without specifically identifying the toxicants. Phase I results are intended as a step in specifically identifying the toxicants but the data generated can also be used to develop treatment methods to remove toxicity without specific identification of toxicants. The presence and the potency of the toxicants in the samples are detected by performing various manipulations on the sample and by using aquatic organisms to track the changes in the toxicity. This toxicity tracking step is the basis of the TIE. Another step, if possible, is to separate the toxicants from the other constituents in the sample in order to simplify the analytical process. Many toxicants must be concentrated for analysis. While this TIE approach was developed for effluents, the methods and techniques have direct applicability to other types of aqueous samples, such as ambient waters, sediment pore waters, sediment elutriates, and hazardous waste leachates. The Phase I manipulations include pH changes along with aeration, filtration, sparging, solid phase extraction, and the addition of chelating (i.e. ethylene-diamine-tetraacetate ligand (EDTA)) and reducing (i.e. sodium thiosulphate) agents. The physical/chemical characteristics of the toxicant are indicated by the results of the toxicity tests conducted on the manipulated samples.

Phase II (EPA, 1989C) describes methods to specifically identify toxicants if they are non-polar organics, ammonia, or metals. This phase is incomplete because methods for other specific groups, such as polar organics, have not yet been developed. Phase III (EPA, 1989D) describes methods to confirm the suspected toxicants. It is applicable whether or not the identification of the toxicants was made using Phase I and II. Complete Phase III confirmations have been limited to date, but avoiding Phase III may invite disaster because the suspected toxicant(s) was not the actual toxicant(s).

Slide 15

Presence of toxic concentrations does NOT prove the cause of toxicity

---

Traditionally, when an effluent has been identified as toxic or is suspected of being toxic to aquatic organisms, a sample of the wastewater is analysed for the 126 "priority pollutants". The concentration of each priority pollutant present in the sample is subsequently compared to literature toxicity data for the pollutant, or is compared to EPA's Ambient Water Quality Criteria or state standards for aquatic life protection for that compound. The goal of this exercise is to determine which pollutants in the wastewater sample are responsible for the effluent toxicity. Unfortunately, determining the source of the effluent's toxicity is rarely straightforward. These 126 chemicals have become known as the "toxic pollutants," conveying an implication that they constitute the universe of toxic chemicals, but the priority pollutants are only a tiny fraction of all chemicals.

Slide 16

Toxicity cannot be "SEEN".

For TIEs, we have learned that this adage is important and not to bias our results with the way something appears or even smells. While clear samples might imply that the effluent is non-toxic, it is our experience that clear samples might actually imply toxicity!

Slide 17

Phase I: Characterisation
- pH adjustment
- filtration

- aeration
- C18 SPE column extraction
- EDTA addition
- sodium thiosulphate addition
- graduated pH

The pH adjustment is done at 3 pHs. These include the initial pH, pH 3 and either pH 9 or pH 11 (acute or chronic, respectively). Samples of the effluent are filtered and aerated and then tested. Aliquots are extracted on a solid phase extraction (C18) column which is eluted with methanol and tested, the post-column effluent is also tested. Additionally, EDTA and sodium thiosulphate are added to aliquots of the effluent to complex cationic metals and reduce oxidants, respectively. Finally, a graduated pH test is done to evaluate the effect of ammonia or pH sensitive toxicants. Typically with these tests, it is possible to evaluate whether polymers, surfactants, non-poplar organics, cationic metals, chlorine, and ammonia are present.

Slide 18

Concentrate on the toxicity characterisation steps that are the most clear-cut and have a major effect on the unidentified toxicants

It is important for the TIE that the results from Phase I are used to evaluate which manipulations gave the most clear information. The results of acute toxicity tests for *Ceriodaphnia dubia* are shown in Slide 19. If, for example, it was found that the toxicity was reduced by both the addition of EDTA and by the addition of sodium thiosulphate, then this table could be used to indicate the next logical step in the process.

Slide 19

	EDTA DOES remove toxicity	**EDTA DOES NOT remove toxicity**
Thiosulphate DOES remove toxicity	Cd Cu Hg	Ag Se (Selenate)
Thiosulphate DOES NOT remove toxicity	Zn Mn Pb Ni	Fe Cr(III) Cr(VI) As (ite or ate) Se (selenite) Al

Slide 20

For non-polar organics, TIE procedure basically relies on two principles: Concentrate and Separate

---

Removal of the toxicity by a C18 column in Phase I indicates that an organic compound(s) might be the cause of the toxicity. Recovery of the methanol eluate can be further analysed. For this step, a gradient of water/methanol solvent that elutes material from the octadecyl solid phase column by changing the polarity of the solvent can be used. Eight fractions are used: 25, 50, 75, 80, 85 90, 95 and 100% methanol/water. Typically, for pesticides the toxicity elutes in the 75-90% fractions. These can then be analysed using the procedures shown in the next slide.

Slide 21

Phase II: Identification

Non-polar organics:	SPE, HPLC, GC/MS, MS-MS
Cationic metals:	AA, ICP, ICP-MS
Volatiles:	Purge and Trap GC/MS
Ammonia, cyanide, sulphide:	Specific ion/colorimetric electrodes
Polar organics:	LC-MS
Anionic metals:	AA, ICP, ICP-MS

The various analytical tools available to measure the suspect toxicants are shown in this slide. At this time we have not observed any volatile compounds in effluents and we have very little experience with the polar organics. Otherwise the analytical tools are those used by analytical chemists for most analyses.

Slide 22

Also found that the C18 SPE column used for non-polar compounds removes:

- nickel, zinc, silver, copper (erratic) and aluminium

However, the use of the C18 column for effluents and water samples is not always clear-cut. In fact, it is surprising that the column can also retain some metals which in turn would not be eluted by methanol.

Slide 23

Phase III: Confirmation

- mass balance
- correlation

- different effluent samples over time
- test with other species
- spike effluent with suspect toxicants
- consider bioavailability

This phase of the TIE consists of a number of steps intended to confirm that the suspect cause(s) of toxicity is correctly identified and that all the toxicity is accounted for. Typically, this confirmation step follows experiments from the toxicity characterisation step (Phase I) and analysis and additional experiments conducted in toxicity identification (Phase II). However, there may often be no identifiable boundary between phases. In fact, all three phases might be underway concurrently with each effluent sample and depending on the results of Phase I characterisation, the Phase II identification and Phase III confirmation activities might begin with the first sample evaluated. Rarely does one step or one test prove conclusively the cause of toxicity in Phase III. Rather, all practical approaches are used to provide the weight of evidence that the cause of toxicity has been identified. The various approaches that are often useful in providing that weight of evidence consist of correlation, observation of symptoms, relative species sensitivity, spiking, mass balance estimates and various adjustments of water quality.

Confirmation is important to provide data to prove that the suspect toxicant(s) is the cause of toxicity in a series of samples and to assure that all other toxicants, that might occur in any sample over time, are identified. There may be a tendency to assume that toxicity is always caused by the same constituents, and if this assumption carries over into the data interpretation but the assumption is false, erroneous conclusions might be reached.

Slide 24

Chemicals identified as causing the toxicity include:

zinc	salinity	ammonia
nickel	diazinon	malathion
chlorfenvinphos	dichlorvos	carbofuran
methyl parathion	chlorine	surfactants

The US EPA has tracked the results of the TIEs conducted at their Duluth laboratory over a number of years. For 75 effluents consisting of 34 industrial discharges, 36 municipal discharges, and 7 from other sources (ambient waters, elutriates, or hazardous wastes) a variety of chemicals that caused the toxicity have been identified.

Slide 25

SUMMARY

- Toxicity cannot be "SEEN"
- US EPA's program for controlling the discharge of toxic chemicals is authorised by Federal law and regulation.
- Toxicity tests are part of the NPDES permit process.
- TIEs are part of the permit or enforcement process. TIEs may result in changing the wastewater treatment.
- TIEs have identified chemicals of concern that were not identified by single chemical analysis of wastewater.
- **"Discharge into the waters of the United States is a privilege, not a right."**

References

1 Federal Register. 1984. US EPA: Development of Water Quality Based Permit Limitations for Toxic Pollutants; National Policy. EPA, Volume 49, No. 48, Friday 9 March 1984.
2 US EPA. 1985. Methods for Measuring the Acute Toxicity of Effluents to Freshwater and Marine Organisms. 3rd Edition. EPA-600/4-85/013. Environmental Monitoring and Support Laboratory, Cincinnati, OH.
3 US EPA. 1989A. Toxicity Reduction Evaluation Protocol for Municipal Wastewater Treatment Plants. EPA/600/2-88/062. Water Engineering Research Laboratory, Cincinnati, OH.
4 US EPA. 1989B. Generalised Methodology for Conducting Industrial Toxicity Reduction Evaluations (TREs). EPA/600/2-88/070. Water Engineering Research Laboratory, Cincinnati, OH.
5 US EPA. 1989C. Methods for Aquatic Toxicity Identification Evaluations: Phase II Toxicity Identification Procedures. EPA/600/3-88/035. Environmental Research Laboratory, Duluth, MN.
6 US EPA. 1989D. Methods for Aquatic Toxicity Identifications Evaluations: Phase III Toxicity Confirmation Procedures. EPA/600/3-88/036. Environmental Research Laboratory, Duluth, MN.
7 US EPA. 1989E. Short-term Methods for Estimating the Chronic Toxicity of Effluents and Receiving Waters to Freshwater Organisms. Second Edition, EPA/600/4-89/001 and supplement EPA/600/4-89/001A. Environmental Monitoring and Support Laboratory, Cincinnati, OH.
8 US EPA. 1991. Technical Support Document for Water Quality-based Toxics Control. EPA/505/2-90-001. Office of Water, Washington, DC.
9 EPA. 1991. Methods for Measuring the Acute Toxicity of Effluents to Freshwater and Marine Organisms. 4th Edition. EPA-600/0-00/000. Environmental Monitoring and Support Laboratory, Cincinnati, OH.
10 US EPA, 1991. Methods for Aquatic Toxicity Identification Evaluations: Phase I Toxicity Characterisation Procedures. EPA-600/6-91/003. US Environmental Protection Agency, Duluth, MN. T. J. Norberg-King, D. I. Mount, E. J. Durhan,

G. T. Ankley, L. P. Burkhard, J. Amato, M. Lukasewycz, M. Schubauer-Berigan and L. Anderson-Carnahan, Eds.

11 US EPA, 1993A. Methods for Aquatic Toxicity Identification Evaluations: Phase II Toxicity Identification Procedures for Acutely and Chronically Toxic Samples. EPA-600/R-92/080. US Environmental Protection Agency, Duluth, MN. E. J. Durhan, T. J. Norberg-King and L. P. Burkhard, Eds.

12 US EPA, 1993B. Methods for Aquatic Toxicity Identification Evaluations: Phase III Toxicity Identification Procedures for Acutely and Chronically Toxic Samples. EPA-600/R-92/081. US Environmental Protection Agency, Duluth, MN. E. J. Durhan, D. I. Mount and T. J. Norberg-King, Eds.

13 US EPA. 1993C. Methods for Measuring the Acute Toxicity of Effluent and Receiving Water to Freshwater and Marine Organisms. 4th Edition. EPA 600/4-90-027F. Environmental Monitoring Systems Laboratory, Cincinnati, OH.

14 US EPA. 1995A. Short-term Methods for Estimating the Chronic Toxicity of Effluents and Surface Waters to Marine and Estuarine Organisms. 3rd Edition. EPA 600/4-91-002. Environmental Monitoring Systems Laboratory, Cincinnati, OH. D. Klemm, G. Morrison, T. Norberg-King, M. Heber and W. Peltier (Eds).

15 US EPA. 1995B. Short-term Methods for Estimating the Chronic Toxicity of Effluents and Surface Waters to Freshwater Organisms. 2nd Edition. EPA 600/4-91-003. Environmental Monitoring Systems Laboratory, Cincinnati, OH. P. Lewis, D. Klemm, J. Lazorchak, T. Norberg-King, M. Heber and W. Peltier (Eds).

The Ecotoxicological Assessment of Hydrocarbons in an Urban Aquatic System

R. H. Jones, D. M. Revitt, R. B. E. Shutes and J. B. Ellis

Urban Pollution Research Centre, Middlesex University, Bounds Green Road, London N11 2NQ, UK

ABSTRACT

The water and sediment concentrations of thirty-nine hydrocarbons have been monitored in the lower, contaminated reaches of an urban catchment in NW London. The same range of aliphatic and polyaromatic hydrocarbons have been measured in the tissues of *Lymnaea peregra* (the wandering snail) which were exposed for periods of up to 32 days at four monitoring sites. The measured tissue levels together with two further variables, mortality and rainfall, have been subjected to Principal Component Analysis to identify the level of association between each measured variable and hence the dependence of invertebrate mortality on specific hydrocarbons or groups of hydrocarbons. This tissue levels of alkanes in *L. peregra* are dominated by the C_{20}-C_{24} range and fluoranthene and pyrene are the dominant polyaromatic hydrocarbons. Statistical analysis identifies important correlations between hydrocarbon tissue levels, particularly for polyaromatic hydrocarbons and their methylated derivatives, with rainfall and organism mortality at two of the sites monitored. However, the hydrocarbon tissue data are difficult to interpret in terms of the original sources of the compounds suggesting that significant changes in the relative compositions of the hydrocarbons occur following bioaccumulation.

1 INTRODUCTION

Oil derived contaminants are recognised as a major constituent of storm-water pollutant loadings in urban areas[1,2] and there are increasing concerns about their potential impacts on receiving water ecosystems.[3] The ability of both aliphatic hydrocarbons and polyaromatic hydrocarbons to disrupt important cell functions[4] and to initiate mutagenic and carcinogenic processes[5] has been shown in laboratory tests. Hydrocarbons may enter the freshwater environment from diagenetic, pyrolytic or biosynthetic sources and, although subjected to chemical and physical weathering and biodegradation processes, will retain identifiable source characterising parameters. These include the chromatographic detection of an unresolved complex mixture (indicating petrogenic pollution), comparison of odd:even carbon chain length alkanes (biosynthetic or petrogenic origin), determination of n-C_{17} to pristane and n-C_{18} to phytane ratios (weathering or degradation) and identification of alkylated polyaromatic hydrocarbons (non-pyrolytic origin).[6]

This paper investigates the extent and impact of hydrocarbon contamination in the waters and sediments of the lower reaches of the Silk Stream catchment located in NW London, UK. This catchment drains a total area of 5239 hectares of which 65% is urbanised resulting in approximately 60% of the annual flow volume being derived from impermeable surfaces.[7] The urban river feeds directly into the Welsh Harp which is a water body categorised as a Site of Special Scientific Interest, because of its ecological status and bird populations, and therefore needs protection from the potentially toxic impacts associated with elevated hydrocarbon concentrations.

The bioaccumulation potential and subsequent toxicity of hydrocarbons is dependent on a range of exogenous and endogenous factors and these are investigated using *Lymnaea peregra*, the wandering snail, as an aquatic indicator organism in *in-situ* toxicity tests. The affinity of hydrocarbons for the lipophilic components of biological membranes and cellular materials is assessed by measuring the total body/tissue burdens of a range of 39 alkanes and polyaromatic hydrocarbons following exposure of this species to polluted urban aquatic environments. These data, together with two further measured variables, mortality and rainfall depth, are subjected to a Principal Component Analysis statistical procedure to deduce the relationships, if any, between hydrocarbon accumulation and lethal toxicity and to examine if these are influenced by changes in river flow which may occur during wet weather conditions.

2 SAMPLING AND ANALYTICAL TECHNIQUES

The locations of the five sampling sites, for which data are presented in this paper, are identified in Figure 1. Site 1 represents a background site, outside the Silk Stream catchment, and provides an abundant source of unpolluted *L. peregra* for relocation to the contaminated urban sites. Site 2 is located approximately 100 m above an oil boom and at this site the bed sediment is composed of consolidated clay with patches of coarse gravel. Sites 3 and 4 are immediately upstream and downstream respectively of the oil boom and both possess fine, highly organic substrates. There are considerably reduced velocities at these sites during dry weather conditions due to canalisation and channel management having taken place. The most downstream site (Site 5) is at the lower end of the Welsh Harp Reservoir and is characterised by a notable

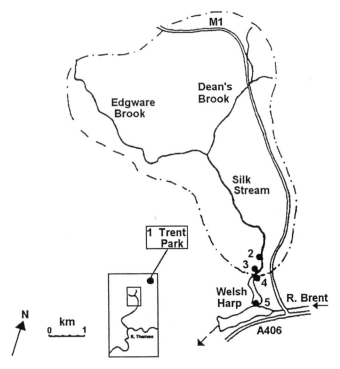

Figure 1 *Sampling Site Locations Within the Silk Stream Catchment*

decrease in sediment organic composition (<20% by loss-on-ignition compared to >60% at Sites 3 and 4).

The collected *L. peregra* were transferred to suspended cages (50-100 organisms per cage) at the four identified locations on the Silk Stream. The cages were kept *in-situ* for periods of between 19 and 32 days and were inspected at weekly or more frequent intervals to ascertain the number of dead organisms and to remove five live animals for hydrocarbon tissue analysis. The organisms were cleaned thoroughly, dried and weighed prior to hydrocarbon extraction with dichloromethane and subsequent analysis by gas chromatography/mass spectrometry in selected ion monitoring mode.

The hydrocarbon tissue concentration results together with measured mortality and rainfall during the exposure periods have been subjected to Principal Component Analysis (PCA) to sort initially correlated data into a hierarchy of statistically independent modes of variation which explain progressively less of the total variation. Hydrocarbon data are suited to this method of analysis because of the correlations between large numbers of individual components of homologous series such as alkanes and closely related compounds within the polyaromatic hydrocarbon group. Similar approaches have been applied to water pollution data but there are few instances of statistical applications of this type to the analysis of macroinvertebrate tissue pollutant concentrations.[9]

3 HYDROCARBON CONCENTRATIONS AND DISTRIBUTION IN WATER AND SEDIMENT SAMPLES

The total alkane and polyaromatic hydrocarbon (PAH) concentrations in the water and sediment phases show the expected increases at the downstream urbanised sites relative to the background site (Table 1). The changes with urbanisation are more elevated for PAHs with comparable increases being observed in both phases. The PAH values recorded in sediments are generally an order of magnitude higher than those previously quoted for freshwater surface sediments[10, 11] although comparable levels have been reported for surface sediments collected from the Great Lakes.[12, 13] The domination of urban runoff contributions to the downstream Silk Stream sites is also indicated by the total alkane levels being approximately ten times higher than those previously reported for other similar catchments.[14, 15]

Comparison of the hydrocarbon levels at the above-and below-boom sites (Sites 3 and 4) suggests that during the sampling period only the water total alkane concentrations were significantly reduced by this device with a maximum removal efficiency of 25%. However, the effective absorption, co-precipitation and sedimentation effects which occur within the Welsh Harp basin for these essentially hydrophobic pollutants are clearly demonstrated by the 50 to 80% reductions recorded in both water and sediment total alkane and PAH levels between Sites 4 and 5.

The alkane distribution profiles are typified by a series of resolved peaks superimposed on a generally small unresolved complex mixture. The water phase alkane concentrations exhibit a generally unimodal profile with the highest concentrations occurring in the C_{18}, C_{19}, C_{20} range. The sediment profiles differ in that the greatest concentrations are found in the higher molecular weight range (C_{25}-C_{27}). The ratios of odd to even carbon number chain length alkanes or Carbon Preference Indices (CPI) can be used to assess the relative contributions of recent biogenic alkane production compared to petroleum derived aliphatic hydrocarbon inputs. Two separate indices for the C_{14}-C_{20} and the C_{20}-C_{32} alkane ranges enable contributions from algae and

Table 1 *Mean Total Alkane and PAH Concentrations in the Water and Sediment During the* In-situ *Toxicity Tests*

Site Code	Alkane concentrations		PAH concentrations	
	Water ($\mu g\ l^{-1}$)	Sediment ($\mu g\ g^{-1}$)	Water ($\mu g\ l^{-1}$)	Sediment ($\mu g\ g^{-1}$)
1	110.3	26.5	3.2	2.8
2	741.6	146.3	97.2	74.8
3	884.3	250.3	140.0	120.6
4	665.4	250.6	128.8	110.9
5	322.7	70.6	36.8	25.9

higher plants to be distinguished. The CPI_{14-20} values were close to unity for all water and sediment samples suggesting that algal inputs are not important relative to anthropogenic sources. The CPI_{20-32} values ranged from 1.27 to 3.40 in water samples and from 1.22 to 2.79 in sediment samples with the highest ratios occurring at Sites 1 and 5 indicating important biogenic contributions from plant material at these sites of lower overall hydrocarbon pollution. The isoprenoid compounds pristane (Pr) and phytane (Ph) may also be used to assess possible hydrocarbon sources and phytane, which has been reported to be only of petrogenic origin, was present in both water and sediment at all the monitored sites.

The extent of hydrocarbon biodegradation in aquatic environments can be assessed by examination of the C_{17}:Pr and C_{18}:Ph ratios since the branched isoprenoids are more resistant to bacterial action than the n-alkanes. The mean C_{17}:Pr and C_{18}:Ph ratios are greater than one at all sites indicating that, in general, biodegradation has not occurred to a significant degree although there is more evidence for this in the sediment compared to the water phase. Low degradation rates may be indicative of recent hydrocarbon inputs with insufficient time for the establishment of substantial bacterial action.

The most abundant PAHs in the sediment and water phases are consistently fluoranthene and pyrene which are known to be predominantly derived from combustion sources as opposed to unburnt petroleum. The presence of high ratios of these PAHs relative to their respective alkyl derivatives also provides evidence for recent combustion derived PAH input.[16] The results of this study show that at all sites, the water and sediment phase levels of parental PAHs predominated over their methylated homologues with the methyl derivatives of fluoranthene and pyrene being completely absent at Sites 1 and 5. This may be interpreted in terms of the dominant inputs being from combustion derived PAHs which mask a lower level fossil fuel (unburnt) contribution. However, the latter becomes proportionately of greater significance at the more polluted downstream sites where contributions from used lubricating oils are becoming evident.

4 *IN-SITU* TOXICITY TESTS AND HYDROCARBON TISSUE BIOACCUMULATION

In the marine environment, investigations of petroleum uptake by two contrasting bivalves, *Macoma balthica* (a deposit feeder) and *Mytilus edulis* (a suspension feeder) have shown clear differences between the uptake and depuration patterns of the two organisms.[17] Mussels have been extensively studied and evidence from scope for growth experiments in laboratory, mesocosm and field conditions suggests that tissue levels of aromatic hydrocarbons (with a molecular weight cut-off below fluoranthene) exert a major toxic effect with an additional contribution from aliphatics of $<C_{11}$.[18] The impact of oil derived pollutants on freshwater species has been less extensively studied although recently Boxall and Maltby[19] have reported on the toxicity to

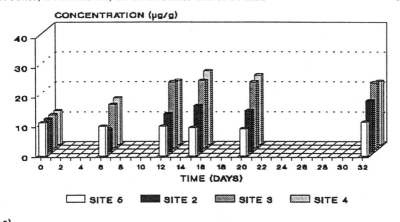

a)

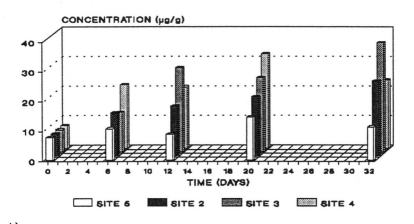

b)

Figure 2 *Temporal and Spatial Variations of a) Total Alkane and b) Total PAH Levels in* L. peregra *During a 32 day Exposure Period*

Gammarus pulex of two to five ring PAHs extracted from sediment contaminated with road runoff. This species was found to be too pollutant sensitive to be used as a biomonitor at the contaminated field sites within the Silk Stream catchment and more resistant species such as *Asellus aquaticus* and *Lymnaea peregra* were therefore employed. The results obtained for *A. aquaticus* have been described previously[20] and this paper examines the hydrocarbon uptake patterns of *L. peregra* and subsequent mortality rates at the four Silk Stream sites (Sites 2, 3, 4 and 5).

4.1 Hydrocarbon Tissue Bioaccumulation

The uptake of total alkanes and PAHs into *L. peregra* tissue during periods of exposure at Sites 2, 3, 4 and 5 are typically represented by the temporal concentration trends shown in Figure 2. Both groups of hydrocarbons demonstrate a marked increase in tissue concentrations during the first three to ten days followed by a tendency for stabilisation of the levels although this is less well defined for the PAHs. The spatial bioaccumulation characteristics clearly mirror the extent of hydrocarbon pollution in the water and sediment at individual sites with Sites 3 and 4 showing little discrimination, particularly for alkanes. The equilibrium tissue values generally approach 30 µg g^{-1} for both alkanes and PAHs at the most contaminated sites and these represent bioconcentration factors of approximately 0.12 and 0.25 relative to sediment concentrations for the alkanes and PAHs, respectively. Considerably higher maximum hydrocarbon accumulation levels have been reported for molluscs in the marine environment with the total saturated hydrocarbon fraction reaching 740 µg g^{-1} and the aromatic fraction achieving values of 290 µg g^{-1}.[21,22]

The individual alkane and isprenoid aliphatic hydrocarbon tissue distributions demonstrate profiles which are dominated by the mid-range C_{20}-C_{24} alkanes and generally reflect those of the surrounding sediment. Investigation of the CPI_{14-20} values shows that there is little or no uptake from biogenic algal sources at any of the sampling locations indicating that C_{14}-C_{20} alkanes accumulated by *L. peregra* are mainly derived from anthropogenic sources. The CPI_{20-32} values for *L. peregra* remain consistently above unity but do not increase significantly throughout individual exposure periods. The uptake of the biogenically derived alkanes is equivalent to that observed in other macroinvertebrates such as *A. aquaticus*[20] and therefore the lower CPI values are most probably explained by increased uptakes of anthropogenically derived compounds by *L. peregra*.

Analysis of the individual PAHs in *L. peregra* tissue shows that phenanthrene, fluoranthene, pyrene and benzo(a)phenanthrene are consistently the major components and typically constitute 35-50% of the total PAH body burden. As was observed in the abiotic phases, fluoranthene and pyrene are the predominant compounds which is consistent with the existence of a large number of diffuse hydrocarbon sources in which combustion sources are prevalent. However, for *L. peregra* the levels of methyl substituted PAHs, particularly methyl-naphthalene, are present at proportionally far higher concentrations in tissues compared to the sediments. This is consistent with the results obtained by the US EPA Mussel Watch Program[23] which found similar disparities between organism hydrocarbon assemblages and the sediments in which they lived. One explanation is that the methyl homologues, which are formed mainly by diagenetic processes, are less tightly particle bound and therefore more readily bioaccumulated in comparison to the parent PAHs which are of pyrolytic origin.

4.2 *In-situ* Toxicity Tests

The mortality rates of *L. peregra* contained in suspended cages at each of the sites were monitored at discrete intervals over each exposure period. The results for a 32 day exposure period are shown in Figure 3 together with the

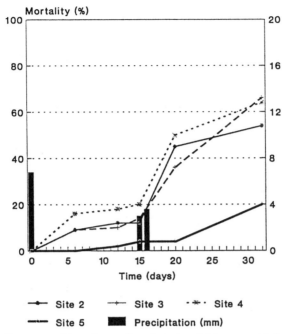

Figure 3 L. peregra *Mortality at Sites 2, 3, 4 and 5 During a 32 day Exposure Period*

measured rainfall data. The spatial hierarchy of mortality rates can be seen to be consistent with the extent of hydrocarbon contamination and hydrocarbon accumulation at each site with the average calculated cumulative mortality rates being 59%, 67%, 71% and 10% at Sites 2, 3, 4 and 5, respectively. The consistently low mortality rate observed at Site 5 would suggest that the experimental cage design is valid in that it imposes low stress on the contained organisms. There is some evidence of a possible relationship between precipitation and mortality (Figure 3) with enhanced mortalities being observed following substantial rainfall which can produce elevated surface pollutant wash-off into the receiving water system. To test the validity of these observations, Principal Component Analysis (PCA) was applied to the collected data for mortality rates, hydrocarbon tissue concentrations and measured rainfall volumes during each exposure period. This statistical procedure identifies the level of association between each measured variable, with the particular aim of identifying specific hydrocarbons or groups of hydrocarbons that are associated with measured *L. peregra* mortality.

Seven principal components (PCs) were identified for the data obtained for Site 2 with the three most important of these explaining 68.3% of the total variance. The most important PC, which explains 34.3% of the total variation (PC1), correlates with a broad range of alkanes (C_{13}-C_{25}) but demonstrates no relationship with either mortality or rainfall. These two parameters are strongly associated with PC2 which is also strongly linked to the methylated homologues of naphthalene, phenanthrene, anthracene, fluoranthene and pyrene. These PAHs suggest a lubricating oil source for this PC which is supported by an additional, although weaker, association with high molecular weight, even carbon numbered alkanes. The high involvement of rainfall in PC2 would indicate that the uptake of hydrocarbons by *L. peregra*, and subsequent mortality, is influenced by oil releases during storm events. PC3 demonstrates a high association with fluorene, fluoranthene, pyrene, benzo(a)anthracene, chrysene, the benzofluoranthenes and benzo(a)pyrene, all of which are characteristic of a combustion derived source. However, the extremely low correlation between this PC and mortality (coefficient; 0.07) reinforces the conclusion that it is the lubricating hydrocarbons and not the combustion source hydrocarbons which exert the greatest toxic threat to *L. peregra* at Site 2.

The increased hydrocarbon contamination which exists at Site 3 does not lead to the existence of clear physical interpretations of the derived statistical relationships. The most predominant PC, which explains 58.6% of the data variance, is associated with a broad mixture of hydrocarbons including the five ringed PAHs, the mid to high range alkanes and the three and four ringed PAHs together with their methyl derivatives. However, mortality is only moderately associated with PC1 (coefficient; 0.34) and rainfall shows an even weaker correlation (coefficient; 0.11). Mortality and rainfall are most strongly associated with PC5 which explains only 3.8% of the variance in the total collected data for Site 3. The only hydrocarbon which influenced PC5 was fluorene and the implication is that at this highly polluted site, hydrocarbons cannot be considered to be strong contributors to toxicity in *L. peregra*. Other factors such as the high sediment oxygen demand (2.57 ± 0.42 g m^2 d^{-1}) and additional non-measured pollutants are believed to be exerting the major toxic effect at this site and the correlation between mortality and rainfall would be consistent with low dissolved oxygen levels as a consequence of remobilisation of disturbed fine sediments during storm conditions.

Mortality is strongly associated with two PCs at Site 4 (PC1 and PC3). PC3 is also highly correlated with rainfall and the predominance of high coefficients for the methylated PAHs implies an important lubricating oil source component. The lower affinity of hydrocarbons from this source with particulates would lead to increased mobility, and consequently bioavailability, during storm events which supports the rainfall involvement in this PC. In contrast PC1, which explains 62.4% of the total data variance shows no association with rainfall but correlates with a wide range of hydrocarbons including C_{18}-C_{24} alkanes and more importantly the full range of monitored PAHs. The results at this site clearly demonstrate the higher *in-situ* toxicities associated with the PAH class of hydrocarbons compared to the aliphatics. The high molecular weight, odd-carbon numbered alkanes such as C_{25} and C_{27} are strongly associated with PC2 and may represent a biogenic source for this PC which is consistent with a very low mortality correlation (coefficient; 0.07).

The greatest spread of the distribution of the system variance was found at Site 5 indicating a 'noisy' data set in which there are a large number of influencing factors despite the lower overall pollution impacts already noted at this site. However, PC1 accounts for a substantial proportion of the system variance (41.5%) and is most strongly related to the high molecular weight alkanes. There is no evidence for a CPI in favour of the odd numbered carbon compounds but the generally low association of the PAHs with PC1 suggests that a biogenic component is likely. The mortality response at Site 5 is highly associated with PC5 which shows no obvious affinity with any groups of hydrocarbons. This is not unexpected given the low recorded mortality rates at this site. The absence of any relationship between rainfall and mortality at Site 5 can be related to the buffering capacity of the lake immediately upstream, as well as the generally reduced contamination both directly from outfalls and from resuspended sediments. The results obtained at Site 5 are in agreement with those for *A. aquaticus*[20] in that associations between tissue hydrocarbons, rainfall and mortality are consistently low.

5 CONCLUSIONS

The measured hydrocarbon levels in the water and sediments at all the monitored sites are explainable in terms of the site locations and extent of urbanisation with the anthropogenic inputs becoming progressively more important at the most polluted sites. However, there is evidence of consistent inputs of hydrocarbons from biogenic sources with carbon preference indices indicating the involvement of higher plants and not algae. The oil boom which is located between Sites 3 and 4 is shown to achieve some reduction in water hydrocarbon levels but produces no reduction in sediment concentrations and does not lead to increased survival rates for the exposed *L. peregra* during the *in-situ* toxicity tests.

The tissue levels of alkanes in *L. peregra* are dominated by the C_{20}-C_{24} range and fluoranthene and pyrene are the dominant PAH compounds with the methyl-substituted derivatives becoming more prominent at the polluted sites. These aromatic compounds exhibit important correlations with organism mortality and rainfall at two sites when the data are subjected to Principal Component Analysis. Although it is generally difficult to interpret the tissue hydrocarbon distributions in terms of the original source of the compound, the statistical analysis clearly implicates the lubricating oil fraction of the hydrocarbon assemblage as an important contributor to organism mortality. Controls of the inputs of hydrocarbon sources of this type are therefore necessary in order to protect the receiving water environment with an emphasis on source control procedures and particularly the education of the public with regard to the use of environmentally sustainable waste disposal methods.

Comparison of hydrocarbon distributions in the biotic and abiotic samples suggests that changes in the relative compositions and modes of variation of the original compounds occur following bioaccumulation by *L. peregra*. This may imply that only a small proportion of the measured hydrocarbons were in the form of partially or non-assimilated compounds such as in the gut content. The lack or scarcity of a mixed function oxidase system in molluscs would allow toxic compounds to accumulate. The more bioavailable compounds, such as the weakly particulate associated methylated PAHs, were able to accumulate to a much greater extent than would be indicated by their levels in the abiotic surroundings. In spite of these characteristics, *L. peregra* is shown to represent a successful biomonitoring species for urban rivers which greatly assists the identification of the fraction of the hydrocarbon assemblage that is primarily responsible for mortality.

References

1. D. B. Porcella and D.L. Sorenson, "Characteristics of Non-Point Source Runoff and its Effects on Stream Ecosystems," Report No. EPA-6003/3-80-032, Oregon, 1980.
2. A. Gavens, D. M. Revitt and J. B. Ellis, "Urban Stormwater Quality, Management and Planning", Water Resources Publications, Colorado, 1982, 79.
3. National Rivers Authority, *The Water Guardian*, January 1995.
4. J. L. Durant, W. G. Thilly, H. F. Hemond and A. L. Lafleur, *Envir. Sci. Technol.*, 1994, **28**, 2033.
5. E. Lavoie, V. Bendenko, N. Hirota, S. Hetch and D. Hoffman, "PAH, Third International Symposium", Ann Arbor, Michigan, 1980, 705.
6. R. H. Jones, D. M. Revitt, R. B. E. Shutes and J. B. Ellis, *Proc. of 6th Int. Conf. on Urban Storm Drainage*, Seapoint Publishing, British Columbia, Canada, 1993, 488.
7. M. J. Hall, "Effects of Urbanization on the Hydrological Regime, Publication 123" IAHS-UNESCO, Amsterdam, 1977, 144.
8. M. T. Bomboi, A Hernandez, F. Marino and E. Honitoria, *Sci. Total Envir.*, 1990, **93**, 523.
9. R. M. Mulliss, D.M. Revitt and R. B. E. Shutes, *Wat. Res.*, 1996, **30**, 1237.
10. B. J. Eadie, P. F. Landrum and W. Faust, *Chemosphere*, 1982, 847.
11. T. S. Bates, P. P. Murphy, H. C. Curl and R. A. Feely, *Envir. Sci. Technol.*, 1987, **21** 193.
12. C. D. Metcalfe, G. C. Balch, V. W. Cairns, J. D. Fitzsimons and B. P. Dunn, *Sci. Total Envir.*, 1990, **94**, 125.
13. D. R. Plowchalk and S. J. Zagorski, *Proc. of the Pennsylvania Academy of Science*, 1986, **60**, 174.
14. J. M. Teal and J. W. Farrington, Rapports et Proces-Verbaux des Reunions Conseil Int. pour L'Exploration de la Mer, 1977, **171**, 79.
15. S. G. Wakeham and J. W. Farrington, "Contaminants and Sediments, Vol.1", Ann Arbor, Michigan, 1980, 3.
16. M. Heit, Y. Tan, C. Klusek and J. C. Burke, *Wat. Air Soil Pollut.*, 1981, **15**, 441.
17. P. D. Boehm, J. Barak, D. Fiest and A. Elskus, "The Tsesis Oil Spill", DOC/NOAA, 1980.
18. J. Widdows and P. Donkin, "The mussel *Mytilus*," Elsevier, Amsterdam, 1991, 383.
19. A. B. A.Boxall and L Maltby, *Wat. Res.*, 1995, **29**(9), 2043.
20. D. M.Revitt, R. B. E. Shutes, R. H. Jones and J. B. Ellis, *Proc. of 7th Int. Conf on Urban Storm Drainage*, 1996 (In Press).
21. R. W. Riseborough, B. E. de Lappe, W. Walker, B. R. T. Simoneit, J. Grimalt, J. Albaiges, J. A. C. Reguiro, A. B. I. Nolla and M. M. Fernandez, *Mar. Pollut. Bull.*, 1983, **14**, 181.

22 L. H. Disalvo, H. E. Guard and L. Hunter, *Envir. Sci. Technol.*, 1975, **9**, 247.
23 J. W. Farrington, R. W. Riseborough, P. L. Parker, A. C. Davis, B. de Lappe, J. K. Winters, D. Boatwright and N. M. Frew, *Woods Hole Oceanographic Institute Technical Report*, WHOI-82-42, 1982.

Selection of Toxicity Assays for Multiple Test Protocols Used in Wastewater Quality Control and Treatment Optimisation

P. Hiley

Yorkshire Water Services Ltd, Knostrop STW, Knowsthorpe Lane, Leeds LS9 0PJ, UK

ABSTRACT

Toxicity tests may help to safeguard the performance of wastewater treatment works, to monitor pollution incidents and assure environmental quality. Tests with organisms representative of the treatment processes and receiving environments may predict damage. No single organism is sensitive to all potential toxicants so a range of tests is needed. The toxic qualities of the inputs and outputs to these environments can be specified to enable the maintenance of a "satisfactory" condition.

The interpretation of potential effects on the treatment system/receiving watercourse requires understanding of the various modes of action of toxic substances. Estimates of total effect cannot realistically take degradation of toxicants during treatment into account. Despite many shortcomings, tests selected from those existing today can be used systematically in an economical, stepwise investigative method. There is a particular need for faster tests to cover a greater range of potential toxicities and at higher sensitivity.

1 OBJECTIVES

To describe the process of choosing, developing and scientifically justifying a toxicity testing suite appropriate to the needs of Water Companies and others employing waste treatment systems, in order to establish the extent of toxicity problems, control inhibition and assess likely effects on receiving ecosystems. To make use of existing tests and identify development needs. "We should tackle today's problems with today's methods".[1]

The protocol and tests described in this paper do not represent the choice of protocols, tests and criteria which Yorkshire Water may have in use now. No recommendations are made for standard values, or for specific tests which may be used in legislative control.

2 INTRODUCTION

By observing the effects of toxic wastes on natural communities we can often determine the approximate source and type of the pollutant. In wastewater treatment we can infer likely consequences on performance resultant from loss of organisms due to toxicity. Using similar organisms in laboratory tests can give a faster, potentially predictive response. Tests may also guide disposal decisions during pollution incidents. Ecosystem data indicated some significant problems with toxic wastewaters in Yorkshire and the existing laboratory tests were used to help understand the magnitude and nature of the toxicity problems facing the Company.

The work concentrated on the toxic properties of whole waste and treated effluent rather than individual components, accepting that the observations may be the result of chemical interactions, synergy, antagony etc. between the various substances present. There has been an English law proscribing the discharge of toxic/inhibitory substances to sewers since 1936[2] though means to enforce it effectively have not been available until comparatively recently.

3 DEFINITIONS

Treatable = degrades to simple components e.g. water and CO_2; *Untreatable* = does not degrade; *Toxic* = harmful to organisms, adversely influences the breakdown of other substances; *Poorly treatable* = degrades slowly, has been used as a euphemism for toxic; *Acclimatise* = become accustomed to a toxicant and may degrade it; *Inhibition* = the degree to which organisms or treatment processes suffer a reduction in their performance. Although toxicity and inhibition have similar definitions, inhibition is of particular relevance to wastewater treatment where it indicates the extent of an effect.

4 IMPACTS OF TOXIC WASTES

4.1 Direct Impacts on Wastewater Treatment

Impacts on the main treatment organisms may be measured by, for example, the inhibition of bacterial respiration. This should be referred to the performance of the works. The "inhibition load" expresses the % inhibition of treatment (e.g. respiration) as if it were extra organic load. The % inhibition of nitrification indicates by how much the works will need to be enlarged. A "toxicity factor"[3] expresses the amount of a wastewater treatment works' under-performance. The "toxicity ratio"[4] ranks components of the sewage in terms of their relative effects on wastewater treatment organisms. Most of the toxics in Yorkshire's wastewaters were degradable at various rates during wastewater treatment.[5] The earlier that toxics degrade in treatment, the less overall effect on performance is likely. The maximum potential impact of toxics would be if they degraded in the final moment of biological treatment, and this is the only consistent estimate of effect which can be presented, since there is no evidence on which an average can be based. The effects of toxic discharges can persist for varying times depending upon the systems impacted. While bacteria may recover from a one-off toxic event in a few hours, grazing invertebrates may take months to re-establish.

4.2 Indirect Effects

1. The effects of a toxic substance on wastewater treatment may appear indirectly, as in the following three examples:
    a) Inhibition of organic decomposers may lead to overgrowth of nitrifying bacteria causing nitrification failure though the nitrification test would show no inhibition.
    b) A single day's discharge of insecticides may destroy the grazing fauna on a percolating filter causing process failure several months later due to excessive biofilm accumulation.
    c) The presence of toxic "uncouplers" may result in increased respiration of bacteria in the three hour test giving the impression of good health, when in fact the bacteria are dying.

4.3 Acclimatisation

Acclimatisation is a general term for the way that wastewater treatment biological systems adapt themselves to the presence of unusual, difficult to treat, and toxic substances.[6] The circumstances in which it develops and is sustained may be outside the control of a Water Company and the treatment works may perform less reliably in consequence. Treatment of a new waste may not take place efficiently until the acclimatisation process is complete. A "memory" of acclimatisation to some types of waste may remain for two or three weeks in long-lived treatment organisms, permitting adaptation to intermittent discharges, whereas for other types of waste a continuous discharge is essential if acclimatisation is to be maintained.[6] Though synergistic effects or cometabolism[7] may increase efficiency with respect to poorly treatable waste components, overall efficiency may be reduced. Some substances can interfere with the maintenance of acclimatisation, for example the treatment of thiocyanate is reduced if phenol, cyanide or ammonia are present.[8,9]

UK regulation of effluent discharges to sewage works and watercourses is based on the BOD test, which uses unacclimatised seed.

5 PREDICTING AND PREVENTING TOXIC IMPACTS

5.1 Field Tests

An impact has to occur for it to be detected, and this may be unacceptable other than in an on-line monitor which facilitates preventative measures. Observations of changes in the communities of naturally occurring organisms are clearly *post-hoc* and relatively insensitive, since gross changes in ecosystem structures are normally monitored weeks or months after they have occurred. The scope for growth test, originally used in marine situations,[10] could be more sensitive as it detects inhibition of growth (or other physiological parameters) of caged invertebrates held within a river or effluent stream.

5.2 Laboratory Tests

Testing a waste in the laboratory enables prediction and possible minimisation of effects, provided they are related back to the field situation. There are difficulties in determining no observed effect concentrations directly due to the length of time the tests take. Relationships between laboratory tests and field observations have been made for high levels of toxicity,[3] and approximate extrapolations may be made to lower levels. 1% of the EC50 concentration was found to be generally a no-effect level.[11] Where the components of a waste have chemical identities, the no-effect values of each component can be established directly with specialised tests. Simulations of wastewater treatment such as porous pot and rolling tubes estimate the treatability of the wastes in admixture with sewage, but acclimatisation makes them more suited to industrial pre-treatment where a there is a more consistent waste quality. The toxicity removal test[5] identifies the elimination of toxic substances in a waste using unacclimatised organisms.

The various qualities of a wastewater are not reliably interrelated, so it is not possible to predict, e.g. % respiration inhibition from the COD:BOD ratio, or nitrification inhibition from the *Photobacterium phosphoreum* assay except on a specific batch of wastewater.[5]

6 SELECTING TESTS FOR SCREENING SUITE

The screening suite should enable the prediction of harmful impacts on the receiving ecosystems, monitoring of wastewater quality and detection of the sources of toxicity. Existing tests have a wide range of reporting times and relative sensitivities due to the nature of the organisms used and the responses measured. All are insensitive to a proportion of the potential toxicants found in wastewater.

6.1 What Constitutes a Good Test?

Standardised and reliable tests strengthen the scientific basis of environmental impact assessments.[1] A good test should have:

National or international protocols, with defined responses to known toxicants,
Organisms relevant to wastewater treatment systems and receiving watercourses,
Relevance to aspects of treatment process performance,
Results which are easy to understand with clear no-effect values,
Protocols close to actual treatment plant conditions,
Precision and reproducibility within given limits,
Machines and materials available so that anyone can repeat the tests,
Results normally produced within 0 - 24 hours,
Ability to detect a large range of potential toxicants with a small range of tests.

Tests tend to be better in some respects than others, e.g. a precise test with a wide detection range for potential toxicants uses a marine organism unrepresentative of wastewater treatment. Only the fastest of tests, reporting generally in minutes, is useful for an on-line role, whereas tests of several months duration may be useful off-line e.g. the plant growth test[12] for checking sludge stockpiles prior to agricultural use. As test variance increases, so the need for understanding and interpreting the results increases. A *Daphnia* result 25% less than another appears "obviously" different yet there may be no statistical justification for this assumption. A non-standard species may be used in preliminary screening (e.g. *Daphnia obtusa* is more sensitive than *D. magna* to some toxicants in crude sewage).

6.2 Specificity of Tests

While there are general toxicants, such as cyanides, chlorine and 3,5-dichlorophenol, many substances target specific groups of organisms (they may have been designed to do this). Synthetic pyrethroids are very toxic to insects and crustacea, but not to bacteria, whereas commercial bactericides are relatively non-toxic to insects and crustacea. Some herbicides target a narrow range of algae and plant species. Low concentrations of some herbicides can give enhanced growth in phytotoxicity tests.

6.3 Adjustments to Samples

Although it may seem to be useful to reduce the samples' toxicity from well known sources e.g. out-of-range pH, high ammonia, etc., correction may alter the nature of other waste components. Therefore it is preferable to subtract the toxicity due to measurable toxicants from the toxicity of the unadjusted sample.

6.4 Controls

Responses to toxicants can vary with genetic differences between strains of test organisms, with conditions prior to the tests, unexpected reactions with media, etc.. Controls and standards help establish to what extent the individual test responses comply with previous experience and influence the credibility of the results. Three or more controls should be set (these relate to sewage treatment): Non-toxic medium = blank control; domestic (satisfactory) sewage or sludge = comparative control; internationally agreed standard toxicants = toxic controls. Domestic (satisfactory) sewage is non-toxic to the respiration inhibition test, whereas it has a modest toxicity to the *P. phosphoreum* assay. Pure ingredients (sand, peat, perlite and adequate nutrients) provide blank results in root elongation and plant growth trials, but comparative controls with domestic sludges give enhanced growth.[12] Where a waste component is known to have a toxic response related to concentration, (e.g. detergents in the *P. phosphoreum* assay) it could either be added in known amounts to a blank or an allowance made for its concentration in the calculation of the test result. It may be seen as a handicap that there is no clear definition of "domestic sewage" or "domestic sludge" in chemical terms; the materials being defined only by the lack of industry in the wastewater treatment works' catchment.

6.5 Data Presentation

Data need to be clear and harmonious so that results from one test may readily be compared to those from another. To avoid the current confusion (e.g. 1% EC50 is the same as 100 times dilution, 5% equivalent to 20 times dilution etc.), as toxicity increases the numbers should always increase in size therefore times dilution and % inhibition are preferred. If a sample is non-toxic the result should be either zero or some low number. Standard values representative of acceptable conditions should be set for tests representative of the receiving ecosystem. These may therefore differ substantially from place to place and time to time.

Chosen tests can be organised to maximise output information and minimise input effort. Table 1 is an example of an economical stepwise investigative method in which the fastest and generally cheapest tests are used first, moving on to more expensive and more precise tests if necessary. Further tests may be selected to replace or complement these, using the criteria given in 6.1 as well as the information at the head of each stage in Table 1. The location of a satisfactory accept/reject decision within Table 1 would depend upon the context and resources available.

The results of the tests require skilled judgement to decide whether or not a "fail" merits further testing. The procedure is context specific, taking into account for example that activated sludge works do not require protection against the effects of insecticides, though the receiving watercourse may need consideration in this respect.

Table 1 Example of a Multiple Test Protocol for Screening Wastewaters in Freshwater Systems

Stage	Examples of tests appropriate to stage
1	Literature search on declared components of the waste.
2	Rapid tests (less than 1 hour) – *P. phosphoreum* toxicity fctor (tf), rapid algae, *Daphnia* contact.
3	1 hour tests – *P. phosphoreum* EC50, *Daphnia* tf, Protozoa tf.
4	1–24 hour tests – *Daphnia* EC50, toxicity removal, *Pseudomonas putida* growth, respiration and nitrification inhibition.
5	1–7 day tests – root elongation, *Lemna* growth, anaerobic inhibition
6	Simulations etc. – porous pot, rolling tube, minidics, *Daphnia* life cycle, microcosms.

At each stage decide to accept, reject or conduct further tests

7 CONCLUSIONS

Laboratory toxicity tests can assess potential for damage, whereas field tests assess damage which has already occurred. The laboratory tests existing at any time can be assembled into a suite which enables economic screening of wastewaters, and tests can be developed to supplement shortfalls in the suite. Particular gaps that were identified were for fast phytotoxicity tests and improvements in precision of the respiration/nitrification inhibition tests. Standards against which to judge the results can be set for no-effect or a given degree of effect, so the suite is adaptable to different circumstances.

Organisms used should be representative of the impacted system and should be unacclimatised to the wastewater under test. They should be capable of detecting all the types of toxicity which may have an impact.

It is possible to effect monitoring and control of toxicity from source to ultimate breakdown using a combination of field observations, field tests and laboratory tests.

References

1. R. M. Robinson (1989). *Hydrobiologia*, **188/189**: 137-142
2. Public Health (drainage of trade premises)Act 1936, HMSO, London, section 26(b) and section 27(1)(a)
3. R. J. Robinson (1988). *Toxicity Assessment*, **3**: 17-31
4. D. Fearnside and H. Booker (1995). "The effects of industrial wastes on nitrification processes found within wastewater treatment works." in: Environmental Toxicology Assessment, ed M. Richardson; Taylor and Francis, ISBN 0 7484 0305 1, pp 281-293.
5. D. Fearnside and P. D. Hiley (1993). "The role of Microtox in the detection and control of toxic trade effluents and spillages." in "Ecotoxicology Monitoring" ed Richardson M, VCH Cambridge, ISBN 3-527-28560-1, pp 319-334,

6. P. D. Hiley (1995) " The economic implications for wastewater treatment of toxic or inhibitory sewage" in: Environmental Toxicology Assessment, ed Mervyn Richardson, Taylor and Francis, ISBN 0 7484 0305 1, pp 387-403.
7. D. Liu, J. F Maguire, G. Pacepavicius and B. J. Dutka (1991). *Environmental Toxicology and Water Quality*, **6**: 85-91
8. A. G. Ashmore, J. R. Catchpole, R. L. Cooper, (1967). *Water Research*, **1**: 605-624
9. *Ibid* (1968). *Water Research*, **2**: 555-562
10. J. Widdows, K. A. Burns, N. R. Menon, D. S. Page and S. Soria, (1990). *J. Exp. Mar. Biol. Ecol.* **138**: 99-117.
11. E. F. King and H. A. Painter (1986). *Toxicity Assessment*, **1**: 27-39
12. P. D. Hiley and B. Metcalfe (1994). "The Yorkshire Water standard plant growth trial for toxicity testing of soils, sludge and sediments", and P. D. Hiley, "The use of barley root elongation in the toxicity testing of sediments, sludges and sewages." in: M. H. Donker, H. Eijsackers, and F. Heimbach "Ecotoxicology of soil organisms" Lewis Publishers, ISBN 0-87371-530-6; pp 179-198.

The Use of Toxicity Identification Evaluation on Industrial Effluent Discharges

V. T. Coombe, K. W. Moore and M. J. Hutchings

Brixham Environmental Laboratory, Zeneca Ltd., Freshwater Quarry, Brixham TQ5 8BA, UK

ABSTRACT

Toxicity Identification Evaluation (TIE) has been developed in the USA as an integral part of the concept of Toxicity Reduction Evaluations (TRE). Such effluent evaluations being required as a result of enforcement action or as a condition of a National Pollution Discharge Elimination System (NPDES) permit. The proposed use of Toxicity Based Consents (TBCs) in the UK has lead to an increasing need for whole effluent toxicity characterisation procedures and the adoption of the TIE/TRE approach seems justifiable. While the TIE concepts were developed for effluents, the methods and techniques have direct applicability to other types of aqueous samples, such as ambient waters, sediment pore waters, sediment elutriates and hazardous waste leachates. The TIE approach is intended to aid those who need to characterise, identify or confirm the cause of toxicity in complex effluents and other aqueous media.

This paper presents the results of TIE investigations on the biotreated effluents from a typical large chemical manufacturing site. The effluents investigated are discharged to the marine environment and the development of suitable marine TIE methodology is described. Several organisms have been identified for control of discharges to surface waters by TBC, including algae, fish and invertebrates. Investigations were carried out using the potential regulatory organism, Pacific oyster (*Crassostrea gigas*). This was found to be most sensitive to the effluents under investigation.

The use of both freshwater and marine invertebrate and algal species in the conduct of TIE/TRE studies is discussed.

1 INTRODUCTION

The use of the assessment of the toxicity of liquid effluent discharges to the environment from both municipal and industrial sources, is set to increase. Traditionally, control of such discharges has been regulated on the basis of physico-chemical parameters such as chemical or biochemical oxygen demand, suspended solids and the quantitative measurement of specific determinands. These measurements have been carried out primarily with the aim of controlling components considered to be particularly damaging to the environment such as highly active or long lived pesticides, chlorinated organics and heavy metals. Compliance with statutory, advised or proposed Environmental Quality Standard (EQS) values has been extensively used as a yardstick, by the regulatory authorities, when determining permitted levels of discharge of the determinands covered. Whilst this system has its advantages in that analyses of individual components are covered by well defined and widely applicable measurement techniques, it has been recognised for some time that EQS values will only ever exist for a tiny proportion of substances used or discharged into the environment and that the scope for increasing the number of substances covered by EQS limits is intrinsically limited by the increasing complexity and cost of adding more and more substances to the list.

Direct Toxicity Assessment (DTA) is seen as one method for a global parametric control of complex effluent discharges. Rather than the measurement of one or more determinands in an effluent that may contain many hundreds or even thousands of substances, an assessment is made of the overall, composite toxicity of the discharge, using defined organisms and protocols, such that an analysis can be made of the likely direct impact of the discharge on the receiving environment. Such a system itself has limitations. Short term toxicity tests are unlikely to detect effects such as bioaccumulation or bioconcentration and, in general, toxicity tests are more complex and expensive than individual determinand analysis. However, direct measurement of toxicity is seen to encompass the combined effect of all the materials present in a discharge, including any synergistic or antagonistic interactions between the components present.

Over the past two years the National Rivers Authority (NRA), now part of the new Environment Agency (EA) have published three information leaflets[1-3] containing updates on research and development activities in the area of aquatic toxicity control and assessment. An inter-laboratory ring test[4] involving the use of different organisms and protocols for toxicity assessment has been conducted and a pilot scheme involving collaboration with a number of dischargers has been undertaken, to assess the benefits and drawbacks of proposed methods of assessment.

An immediate corollary to the introduction of any kind of toxicity based assessment scheme is the question of what to do if a problem is identified with the level of toxicity exhibited by a particular discharge. This paper concentrates mainly on our use of Toxicity Identification Evaluation (TIE) phase I procedures (*vide infra*) in the investigation of the toxicity of a major manufacturing complex situated on a large estuary in Scotland.

2 INITIAL TOXICITY STUDIES

The site chosen for the case study manufactures a wide range of agrochemical and speciality chemical products and intermediates. The site has an 80 year history and effluent disposal has traditionally been handled by estuarial disposal, via long outfall, without prior treatment. The area is one of high natural dispersion but it was recognised that the existing effluent discharge arrangements were unlikely to be sustainable in the long term, particularly under the auspices of Best Available Techniques Not Entailing Excessive Cost (BATNEEC) for effluent disposal, and that work was likely to be necessary in order to meet future consents.

In the early 1990s a commitment was made to investigate impact reduction strategies involving a number of effluent treatment techniques and it quickly became clear that aerobic biotreatment was likely to offer the greatest benefits at the most reasonable cost. After laboratory studies, a pilot-scale biotreatment plant was constructed and operated in 1993-94 and, based on the data obtained, a commitment was made to construct a full-scale biotreatment plant. This decision was based primarily on the belief that biotreatment would offer the most cost-effective solution to producing a neutral, low BOD, low residual COD effluent, for final discharge to the estuary.

Whilst the performance of the pilot scale treatment plant was monitored with respect to the usual sanitary control parameters, such as BOD and COD reduction, an additional measurement parameter was also the degree of effluent toxicity reduction afforded by the biotreatment process. A wide range of toxicity tests was carried out on samples of both treated and untreated effluent. Toxicity monitoring of the site effluent has been carried out on a regular basis for a number of years, together with annual biological surveys of the estuary to monitor the environmental impact of the site discharge. In order to monitor the effect of biotreatment on the toxicity of the discharge, a range of organisms was chosen, as there was relatively little experience of the differences in sensitivity, exhibited by various trophic levels, to this discharge.

Preliminary evaluations involved the use of *Photobacterium phosphoreum* (Microtox^{TM}), the marine copepod *Tisbe battagliai*, *Scophthalmus maximus* (turbot), *Oncorhynchus mykiss* (rainbow trout), *Crassostrea gigas* (Pacific oyster) and the marine alga *Skeletonema costatum*.

Pilot-scale biotreatment caused a marked decrease in toxicity when assayed by the Microtox^{TM} test with the mean EC50 concentration, for all samples tested, falling from an average of 12.9% by volume ($^v/_v$) before treatment to 31% $^v/_v$ after treatment. There was no apparent correlation between the toxicity as measured by the Microtox^{TM} test when compared to any of the other test organisms.

Fish proved to be relatively insensitive to toxicity of the biotreated effluent. However, because of their importance in the ecology of the estuarial ecosystem, fish testing has been carried out for a number of years on the site effluent. LC50 concentrations measured using both turbot and rainbow trout for untreated effluent ranged from 20 to 40% $^v/_v$ and from 20% to 75% $^v/_v$ after biotreatment.

Algal tests using *Skeletonema costatum* indicated a greater degree of sensitivity than either fish or Microtox^{TM} with biomass based EC50 concentrations in the range 6% to 13% $^v/_v$ for the range of samples tested. There was no apparent toxicity reduction on biotreatment when assessed using this organism.

Toxicity assessment using Pacific oyster indicated that it was the most sensitive organism of the suite of test organisms used by at least an order of magnitude. EC50 concentrations for untreated effluent were in general < 1% $^v/_v$ and there was again no reduction in toxicity after biotreatment.

Results from the complete suite of tests described above indicated that Pacific oyster was the most sensitive organism. Were the introduction of toxicity based controls to follow the United States Environmental Protection Agency (US EPA) National Pollutant Discharge Elimination System (NPDES) scheme, then Pacific oyster would be the regulatory organism chosen. Consideration of other environmentally relevant information, however, may serve to modify this choice.

3 TOXICITY IDENTIFICATION EVALUATION

When a problem with the toxicity of a discharge is identified, one of the ways of tackling that problem is the initiation of a Toxicity Reduction Evaluation (TRE). This is a relatively complex procedure which involves a number of discrete steps. There is no rigid basis for the precise manner in which such an investigation must be conducted, although extensive protocols for the assessment of the sources of toxicity within a particular discharge have been published by the US EPA.[5-8] In the US, the use of whole effluent toxicity assessment has been part of the control mechanism for the regulation of effluent discharges for a number of years.

A major part of the TRE exercise is the phased investigation of the particular sources of toxicity in a given discharge, the so called Toxicity Identification Evaluation (TIE) which involves three main stages.

a) toxicity characterisation (phase I)
b) toxicity identification (phase II) and
c) toxicity confirmation (phase III)

A particular feature of TIE is that it requires toxicity testing on an unprecedented scale. Typically 200 to 300 individual test samples may be generated for toxicity testing depending on the scope of the TIE and testing, for example, in duplicate obviously doubles this number. This requires that large numbers of toxicity tests be carried out simultaneously and this in turn forces certain requirements of the toxicity tests themselves:

- Sufficient test organisms must be available at the start of the tests for testing to be carried out simultaneously.
- The volume of sample required for the test must be sufficiently small to obviate the need for production of prohibitively large quantities in the TIE manipulations.
- Assessment must be possible in a relatively short time after test completion or there should be a method of preserving the state of the tested organisms at the end of the test period.
- The test must be cost effective and amenable to large scale use in order to avoid prohibitive cost.

Pacific oyster larval toxicity testing is one of the test protocols which fulfils all of the requirements above without significant modification. It also had the major advantage that, since *Crassostrea* was likely to be the regulatory organism of choice, using it as the TIE test organism eliminates interspecies differences in toxicity response.

4 TIE PHASE I RESULTS

Phase I of the TIE procedure was carried out in 1995 using the protocol outlined in Figure 1. The results are presented in Figures 2 to 5.

A relatively small change in toxicity was noted as a result of adjusting the pH to 3 with a larger reduction, amounting to a 40% toxicity removal being

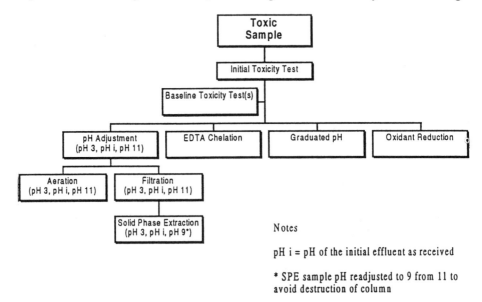

Figure 1 *Phase I Toxicity Characterisation Procedures*

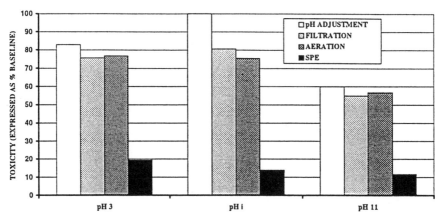

Figure 2 *Phase I Toxicity Tests – pH Adjustment, Filtration, Aeration and Solid Phase Extraction (SPE)*

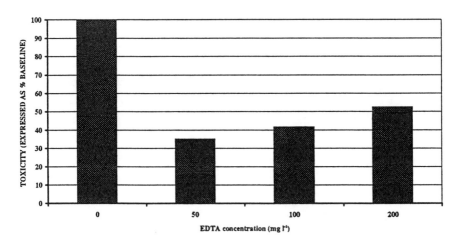

Figure 3 *Phase I Toxicity Test Results – EDTA Chelation*

noted after adjustment to pH 11 (Figure 2). Subsequent filtration and aeration steps produced only small reductions in toxicity. Solid phase extraction (SPE) showed significant effects at all pH values, reducing measured toxicity levels by up to 90%.

Addition of EDTA to samples of the effluent elicited significant toxicity reductions (Figure 3). The largest drop on toxicity was noted at an EDTA concentration of 50 mg l^{-1}. Increasing EDTA levels past 50 mg l^{-1} did not achieve further toxicity reduction, with toxicity subsequently showing a slight rise with increasing EDTA concentration, presumably due to the increasing toxicity of residual EDTA. Such behaviour indicates that a metal is at least partially responsible for toxicity. Subsequent analysis by atomic emission spectroscopy coupled with comparison of known EC50 data for the effects of heavy metals to *Crassostrea gigas*[9-13] suggested that copper was the most likely cause of metal based toxicity.

The relationship between observed toxicity and copper content after various TIE manipulations is illustrated in Figure 4. Copper concentration in general showed good correlation with toxicity. Significantly, most of the copper remaining after filtration at pH 11 is subsequently retained by a solid phase extraction column at pH 9. This suggests that the remaining copper may be complexed in some way with organic components in the effluent.

Figure 5 illustrates the removal of toxicity after filtration by solid phase extraction at pH 3, pH i (pH of the initial sample as received) and pH 11,

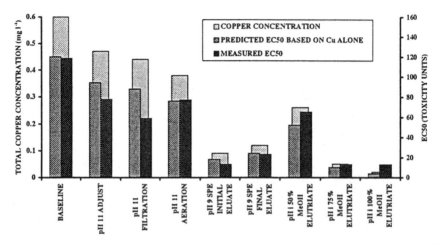

Figure 4   *Phase I Toxicity Test Results Compared with Copper Concentrations and Resultant Predicted Copper Toxicity*

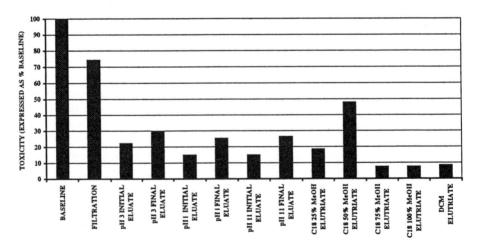

Figure 5   *Phase I Toxicity Test Results – Solid Phase Extraction Column Eluate and Elutriate Toxicity*

together with the recovery of toxicity from solid phase extraction media with increasingly non-polar effluents up to and including the use of dichloromethane (DCM).

5 CONCLUSIONS

The results of the TIE Phase I study clearly indicate that a major source of toxicity in the effluent sample tested is metal (i.e. specifically copper). The results also indicate that there is a further significant source of toxicity, related to a copper containing, non-polar organic fraction. The reduction in the toxicity of the effluent on addition of EDTA is a classic response in the presence of certain dissolved cationic metals. The reduction in effluent toxicity associated with passage of the effluent through the solid phase extraction column may be an expression of toxicity due to the dissociation of copper from organo-copper complexes, an expression of the toxicity of the organo-copper complexes themselves, or the expression of organic component toxicity, unrelated to the copper content of the effluent. Further studies are underway in an attempt to identify these components and to determine what role, if any, the remaining copper plays in the toxicity of the material extracted by the SPE column.

The use of DTA to estimate actual environmental impact, would seem to present an assessment technique more closely related to the final objective of protecting the aquatic environment than bulk sanitary or chemical specific analyses. Pragmatism may, however, limit the regulatory use of DTA to control discharges until the performance of the DTA methods and the correlation of their output with ecological observation are more fully validated.

Where effluent toxicity is recognised to be an issue in environmental management, with or without the presence of regulatory pressure, TRE and TIE procedures have been found to offer a practicable option for a structured response to the issues identified. These investigations can prove complex and extensive however, and will certainly require a case specific approach in order to maximise the environmental benefit and minimise the potentially large resource costs involved.

References

1. The Toxicity-Based Consent - A step towards more effective control of complex effluents within the UK NRA information leaflet No. 1.
2. Direct Toxicity Assessment - A step towards better environmental protection within the UK NRA information leaflet No. 2.
3. Aquatic Toxicity Control and Assessment - An update on current research and development. NRA information leaflet No. 3.
4. P. Whitehouse, P. A. H. van Dijk, P. J. Delaney, B. D. Roddie, C. J. Redshaw and C. Turner. 1996. This volume.
5. EPA (1991). Methods for Aquatic Toxicity Identification Evaluations - Phase I

Toxicity Characterisation Procedures (Second Edition). Eds. T. J. Norberg-King *et al.* EPA/600/6-91/003, Environmental Research Laboratory, US EPA.

6   EPA (1993). Methods for Aquatic Toxicity Identification Evaluations - Phase II Toxicity Identification Procedures for Samples Exhibiting Acute and Chronic Toxicity. Eds. T. J. Norberg-King *et al.* EPA/600/R-92/080, Environmental Research Laboratory, US EPA.

7   EPA (1993). Methods for Aquatic Toxicity Identification Evaluations - Phase III Toxicity Confirmation Procedures for Samples Exhibiting Acute and Chronic Toxicity. Eds. T. J. Norberg-King *et al.* EPA/600/R-92/081, Environmental Research Laboratory, US EPA.

8   Margarete A. Heber and Teresa J. Norberg-King. Editorial synopsis - John Tapp, James Wharfe and Stephen Hunt. 1996. This volume.

9   G. Mance (1987). Pollution threat of heavy metals in aquatic environments. Elsevier Applied Science. Chapter 5: Toxicity of metals to Marine Life, 192-205.

10 C. E. Woelke (1972). Development of a receiving water quality bioassay criterion based on the 48 hour Pacific Oyster, *Crassostrea gigas*, embryo. *Washington Department of Fisheries Technical Reports*, **9**, 1-93.

11 M. Martin, P. B. Osborn, P. Billig and N. Glickstein, (1981). *Marine Pollution Bulletin*, **12**, 9, 305-308.

12 M. P. Coglianese and M. Martin, (1981). *Marine Environmental Research*, **5**, 13-27.

13 M. P. Coglianese (1982). *Archives of Environmental Contamination and Toxicology*, **11**, 297-303.

Overall Cost Savings and Treatment Optimisation by Segregation of the Component Waste Streams in a Toxicity Reduction Evaluation

J. H. H. Looney,[1] M. A. Collins,[2] M. S. Stein[3] and S. R. Harper[3]

[1]Parsons Engineering Science, Newporte House, Low Moor Road, Doddington Road, Lincoln, UK
[2]Parsons Engineering Science, Inc., Two Flint Hill, 10521 Rosehaven Street, Fairfax, Virginia, VA22030, USA
[3]Parsons Engineering Science, Inc., 57 Executive Park South, THE, Suite 590, Atlanta, Georgia, GA30329-2265, USA

ABSTRACT

The Environment Agency are proposing a new approach for regulating discharge consents from industry based on toxicity based consents (TBCs). This will supplement the existing chemical specific approach and will be applied to discharges that are variable, difficult to interpret because of interactions or are known to contain toxic substances. Parsons Engineering Science has worked with the US EPA deriving the Direct Toxicity Approach initially and then the application of this for industry through Toxicity Reduction Evaluations (TREs). This paper gives a brief overview of the proposed Environment Agency protocol for Toxicity Based Consents and describes the TRE approach. It is supported by data from several studies on how this can be a cost effective approach to regulating effluent discharges.

Toxicity treatability studies were performed as part of a toxicity reduction evaluation at a chemical manufacturing facility to evaluate options for the removal of refractory organic compounds identified as causes of toxicity. Presently, waste streams from a number of different chemical manufacturing processes are combined and treated at the facility's biological wastewater treatment plant (bioplant). Initial batch treatability studies indicated that the toxicity problems could be attributed primarily to one of the component waste streams. Granular activated carbon (GAC) treatment of a slip stream from the bioplant effluent was shown to be an effective means of reducing whole effluent toxicity. However, the costs associated with continuous, long-term GAC treatment suggested that further study of more cost-effective toxicity control measures (e.g., source controls and improved biological treatment) was warranted.

Bench-scale, continuous flow treatability studies were initiated to confirm the toxicity sources; determine if problematic organics could be biologically treated by separate, focused treatment of the primary source waste stream; and to develop and evaluate an overall treatment strategy to meet both toxicity and chemical oxygen demand (COD) discharge limitations. Three separate continuous flow reactors were operated. The first reactor served as the control to simulate treatment of the existing bioplant influent, the second was fed with bioplant influent with the problematic waste stream removed, and the third was used to treat only the problematic waste stream.

This study showed that it is possible to achieve improved COD conversion and toxicity removal at the bioplant by implementing an alternative biological treatment strategy. The results indicated that COD conversion and toxicity removal was improved when the problematic waste stream was treated separately from the other major process waste streams. The net result, when the two effluents were recombined, was that both COD and toxicity were lower and met discharge permit limits. The annualised cost of implementing separate biological treatment of the problem stream (with filtration) was estimated to be nearly an order of magnitude less than the GAC slip stream treatment, which has also been shown to remove toxicity effectively.

1 INTRODUCTION

This paper presents the findings of a toxicity treatability study undertaken by Parsons Engineering Science, Inc. (Parsons ES) for a confidential chemical manufacturing client. The study was performed as part of an ongoing toxicity reduction evaluation (TRE) being performed at the facility. A TRE is a site-specific study designed to identify the causative agents of effluent toxicity, isolate the sources of toxicity, and evaluate the effectiveness of toxicity control options. With increasing regulatory commitments and limited resources for regulatory program compliance, it is important for industrial and municipal dischargers to select the most pragmatic and cost-effective solution to their toxicity compliance issues. This paper focuses on the use of a stepwise TRE process and site-specific information to identify and evaluate cost-effective toxicity control options for an industrial facility.

1.1 Regulatory Requirements

The US National Pollutant Discharge Elimination System (NPDES) permit for the subject chemical manufacturing facility requires acute and chronic toxicity testing on the outfall using *Mysidopsis bahia* (mysids) and *Cyprinodon variegatus* (sheepshead minnow). The permit specifies that 24 h acute tests must demonstrate > 50% survival in the 100% effluent. Based on unfavorable results from the acute mysid tests, the facility initiated the TRE process in January 1994. The overall objective of the TRE has been to meet the > 50% survival criterion for the 24 h acute mysid test as specified in the discharge permit.

1.2 Facility Description

The chemical manufacturing facility discharges approximately 9.8 million litres per day into a salt water receiving system. Approximately 50% of the flow consists of non-process wastewater (e.g., cooling tower blowdown) which receives no treatment prior to discharge. The remaining 50% of the flow is the effluent from the facility's biological wastewater treatment plant (bioplant). A total of 13 individual waste streams feed into the bioplant, with four main process waste streams contributing approximately 93% of the total flow. Average bioplant influent chemical oxygen demand (COD) is approximately 3,000 mg l^{-1} and 90% or greater COD conversion is achieved consistently. Unit processes in the bioplant are shown in Figure 1 and include API separation, equalization, single-stage activated sludge treatment, and secondary clarification. The equalised flow from all waste streams is treated in four aeration basins that are currently operated in parallel. This system offers a degree of operational flexibility including the possibility of segregating and treating problematic wastewater separately, rather than combining and treating all waste streams together. Granular activated carbon (GAC) slip stream treatment is also available for COD polishing to ensure compliance during periods of high loadings or plant upsets.

1.3 Summary of TRE

1.3.1 Phase I – Toxicity Identification. A step-wise, phased TRE approach was utilised to obtain information about the causes and sources of effluent toxicity. Information obtained during the first phase indicated that the bioplant effluent was the primary source of effluent toxicity. In addition, evaluations performed during this phase of the study indicated that housekeeping practices and bioplant performance/operations were not contributing to effluent toxicity. Toxicity identification evaluation (TIE) tests and related chemical-specific analyses indicated that a complex mixture of unknown non-polar organic (NPO) compounds were the primary causes of toxicity at the outfall and also in the bioplant effluent. Moreover, since these NPOs were escaping treatment in the bioplant, it was suspected that they were refractory (i.e., resistant to biological treatment).

Due to the complex mixture of unknown NPO compounds present in the effluent, identification of specific compounds causing toxicity did not prove to be technically feasible using TIE procedures and GC/MS and LC/MS

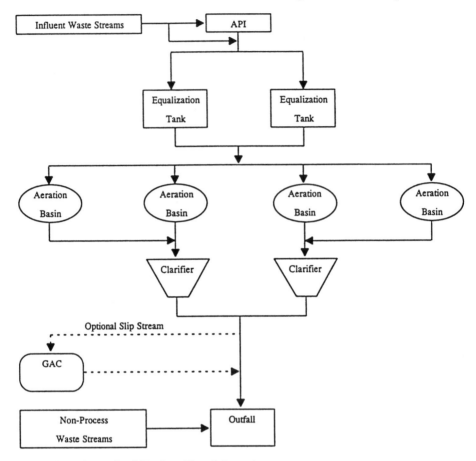

Figure 1 *Generalised Bioplant Flow Schematic*

methods. As a result, it was not possible to identify the sources of effluent toxicity using a chemical-specific approach.

The effectiveness of GAC treatment in removing effluent toxicity was also evaluated during this phase of the study. Bench-scale and full-scale trials indicated that GAC treatment removed acute toxicity from the bioplant effluent. However, the costs associated with continuous, long-term GAC treatment suggested that further study of more cost-effective toxicity control measures (e.g., source controls and improved biological treatment) was warranted.

1.3.2 Phase II – Batch Treatment Tests. Accordingly, toxicity-based source evaluation studies were performed during the second phase of the study to identify the waste streams which contributed to bioplant effluent toxicity. These studies began with an initial toxicity screening, where the potential contribution from each of 13 waste streams discharging to the bioplant was evaluated by measuring the intrinsic toxicity of each using the Microtox™ technique. Based on the results of the screening, individual waste streams were selected for evaluation using refractory toxicity assessment (RTA) testing and "process elimination" tests.

The RTA approach utilised bench-scale, batch biological treatment reactors to evaluate the contribution of selected waste streams to overall toxicity. Sources of toxicity were also evaluated by performing effluent toxicity tests during temporary shutdowns of various manufacturing process units. The results of the screening and RTA tests suggested that although at least three of the four main process waste streams contained refractory toxicity, the toxicity problems could be primarily attributed to just one of these four streams.

1.3.3 Phase III – Continuous Treatment Tests. During the third phase of the study, bench-scale, continuous flow studies were initiated to confirm the toxicity sources; determine if problematic organics could be biologically treated by separate, focused treatment of the primary source waste stream; and to develop and evaluate an overall treatment strategy to meet both toxicity and COD discharge limits. Three separate continuous flow reactors were operated. The first reactor served as the control to simulate treatment of the existing bioplant influent (containing all four major process streams), the second was fed with bioplant influent with the problematic waste stream removed, and the third was used to treat only the problematic waste stream. The results of the third phase of testing are described in more detail in the remainder of this paper.

2 MATERIALS AND METHODS

2.1 Description of Reactors

The study utilised continuous flow, completely mixed, biological treatment reactor systems. Each reactor was constructed of transparent Plexiglas, and held a volume of 14 litres with a sliding baffle that divided the contents into an aeration zone of approximately 9 litres and a clarification zone of approximately 5 litres. All reactors were operated under a laboratory fume hood to minimise fugitive organic vapours in the treatability laboratory. Air was supplied to each reactor by a central air compressor. Air flow was regulated with manually-set rotameters before being passed through a stone diffuser inside each reactor. The diffused air also provided complete mixing within the aeration zone of the reactors.

The premixed influent wastewaters were placed in 36 litre, high-density polyethylene containers equipped with floating lids. Influent wastewaters were replenished on a weekly basis. Peristaltic pumps continuously conveyed the influent to the reactors through plastic Tygon tubing. Effluent continuously flowed from the clarification zone into 18 litre plastic effluent holding containers, which were also fitted with lids to minimise evaporative losses.

The following reactors were operated for the 15-week study period:

- Reactor 1 - treating the four major process waste streams;
- Reactor 2 - treating three major process waste streams without the problematic waste stream; and
- Reactor 3 - treating the problematic waste stream only.

Reactor 1 acted as the control and served as a baseline for comparison with the other reactors. Reactors 2 and 3 were operated to confirm sources of effluent toxicity and to evaluate the possibility of achieving improved toxicity removal using a cost-effective, alternative biological treatment strategy. As shown in Figure 2, the alternative strategy would involve treating the problematic waste stream in one of the existing aeration basins, treating the

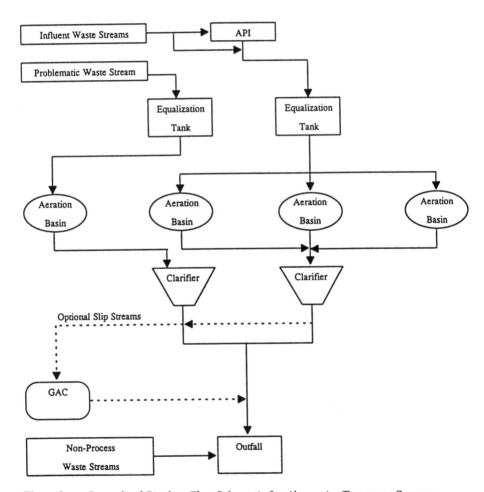

Figure 2 *Generalised Bioplant Flow Schematic for Alternative Treatment Strategy*

Table 1 *Reactor Descriptions and Target Operating Values*

Parameter	Reactor Targets
Aeration Basin Volume (l)	9
Clarifier Volume (l)	5
Flow (l d^{-1})	2.25
Hydraulic Residence Time (days)	4
Sludge Residence Time (days)	40–50
Sludge Wasting Rate (l d^{-1})	0.3
MLSS (mg l^{-1})	4,500
Food/Mass Ratio	0.2
Mixed Liquor pH (su)	8.5
Mixed Liquor DO (mg l^{-1})	>2.5
Aeration Rate (ml min^{-1})	1.5
Effluent Ammonia (mg l^{-1} NH$_3$-N)	0.5–1.0
Effluent Phosphate (mg l^{-1} PO$_4$-P)	3.0–9.0

combined wastewater from all other waste streams in three of the existing basins, and combining the treated effluents prior to discharge. Accordingly, the effluents from Reactors 2 and 3 were tested for toxicity separately, and also combined to reflect conditions that would prevail at the bioplant outfall under the alternative treatment strategy.

2.2 Target Operating Values

Operating parameters for the three reactors are summarised in Table 1. These parameters were based on the prevailing operating strategy at the bioplant.

2.3 Nutrient Addition

All reactors were fed supplemental nitrogen and phosphorous nutrients in the form of ammonium hydroxide (NH$_4$OH) and phosphoric acid (85% solution of H$_3$PO$_4$), to simulate bioplant operation and to ensure that performance was not inhibited because of insufficient nutrients. The nutrients were added to the influent feed containers when new feed was placed in the containers (approximately weekly).

Nutrients were not added to the stock influent samples to minimise the possibility of biological growth during storage. Initially, nitrogen and phosphorous nutrient additions were based on theoretical cell growth requirements. Nutrient doses were adjusted during the course of the study based on reactor effluent data.

2.4 Reactor Operational Monitoring

Reactor performance was monitored by measuring the following parameters at the approximate frequencies indicated:

- Influent total COD - weekly;
- Effluent total and soluble COD - three times per week;
- Aeration basin dissolved oxygen (DO), temperature, and pH - daily;
- Mixed liquor suspended solids (MLSS) - three times per week; and
- Effluent total ammonia as nitrogen (NH_3-N) and effluent total phosphate (PO_4-P) - weekly.

2.5 Toxicity Monitoring

Reactor effluent toxicity was evaluated after the reactors had achieved steady operation by performing 48 h acute toxicity tests with *Mysidopsis bahia* (mysids). The test results were used to calculate both 24 and 48 h LC50 values.

3 RESULTS

3.1 Operational Data

The reactors were operated for a period of 15 weeks from 14 April through 28 July 1995. As shown in Figure 3, overall reactor performance was considered good based on average COD conversion. Average COD conversions of 89% were achieved in Reactors 1 and 2, and 94% conversion was observed in Reactor 3. Average MLSS data are shown in Figure 4. Reactors 1 and 2 exhibited similar average MLSS concentrations of 3,863 and 3,568 mg l^{-1}, while Reactor 3 was operated at a higher average MLSS concentration of 4,330 mg l^{-1}. As shown in Figure 5, average effluent soluble COD was lowest in Reactor 2 (163 mg l^{-1}), followed by Reactor 1 (324 mg l^{-1}), and Reactor 3 (371 mg l^{-1}).

The predicted soluble effluent COD for the alternative biological treatment strategy was calculated based on data from Reactors 2 and 3 and the flow contribution of the individual waste streams (Reactor 2 flow contribution

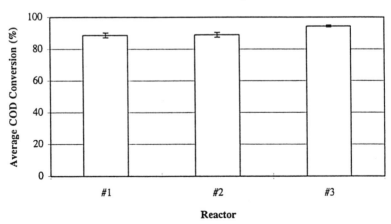

Figure 3 *Average COD Conversion*

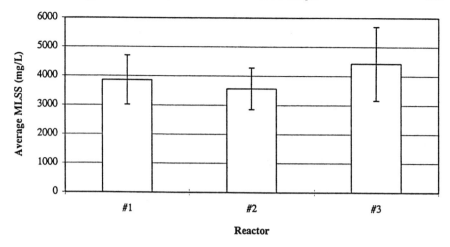

Figure 4   *Average Mixed Liquor Suspended Solids*

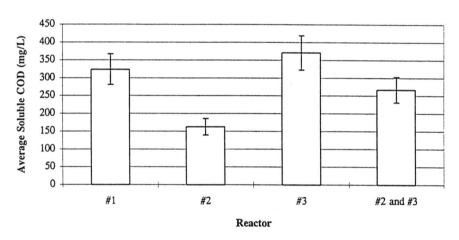

Figure 5   *Average Reactor Effluent Soluble COD*

= 72% and Reactor 3 flow contribution = 28%). As shown in Figure 6, the calculated soluble COD for the combined effluent from Reactors 2 and 3 was consistently lower than that of Reactor 1. These data indicate that better COD conversion was achieved by treating the problematic waste stream separately, rather than treating all of the waste streams together.

3.2 Toxicity Data

3.2.1 Reactor 1. The results of acute mysid toxicity tests performed on the reactor effluent samples are presented in Table 2 and Figure 7. Both 24 and 48 h

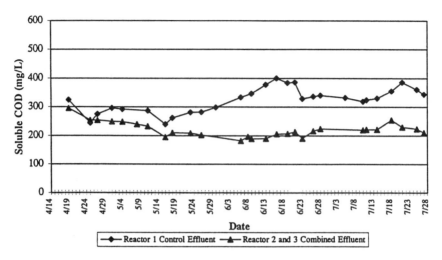

Figure 6 Soluble COD for Combined Effluent

Table 2 Reactor Effluent Acute Mysid Toxicity Data

Reactor	Mysid Toxicity Data			
	24 h LC50 (%)	24 h Toxic Unit	48 h LC50 (%)	48 h Toxic Units
Round 1: 14 June 1995				
Reactor 1–Control	37.3	2.7	13.6	7.4
Reactor 2	>100	<1.0	>100	<1.0
Reactor 3	15.6	6.4	7.4	13.5
Round 2: 7 July 1995				
Reactor 1–Control	57.4	1.7	11.9	8.4
Reactor 2	>100	<1.0	>100	<1.0
Reactor 3	<6.25[a]	>16.0	<6.25[a]	>16.0
Round 3: 13 July 1995				
Reactor 1 - Control	>100	<1.0	43.5	2.3
Reactor 2	>100	<1.0	>100	<1.0
Reactor 3	6.9	14.5	2.3	43.5
Filtered Reactor 3[b]	33.0	3.0	19.0	5.3
Reactor 2 + Reactor 3[c]	37.9	2.6	18.3	5.5
Round 4: 28 July 1995				
Reactor 2	>100	<1.0	>100	<1.0
Reactor 3	<6.25[a]	>16.0	<6.25[a]	>16.0
Filtered Reactor 3[b]	42.0	2.4	15.0	6.7
Reactor 2 + Filtered Reactor 3[c]	>100	<1.0	56.1	1.8

[a] There was <50% survival in 6.25% effluent, which was the lowest concentration tested.
[b] Sample was filtered through standard TSS glass filter prior to testing.
[c] Sample consisted of 72% Reactor 2 effluent and 28% Reactor 3 effluent or filtered Reactor 3 effluent.

LC50 values and their respective toxic units (TU = 100/LC50) are given in Table 2. The 24 h LC50 values for Reactor 1 (control) ranged from 37.3 to (>)100% effluent. The LC50 values obtained for Reactor 1 during Rounds 1 and 2 were similar to values typically obtained for the bioplant effluent (average 24 h LC50 $\sim$40%).

3.2.2 Reactor 2. The 24 and 48 h LC50 values of (>)100% effluent were obtained for Reactor 2 during all four rounds of testing, indicating that the effluent was not acutely toxic to mysids. Comparison of these results to Reactor 1 indicates that removal of the problematic waste stream resulted in a reduction in effluent toxicity. The results confirm previous findings regarding the primary source of bioplant effluent toxicity.

3.2.3 Reactor 3. The 24 h LC50 values for Reactor 3 ranged from 15.6% during Round 1 to <6.25% during Rounds 2 and 4. These results provide additional evidence that this waste stream was the primary source of refractory toxicity. Due to the apparent increase in toxicity from Round 1 to Round 2, an aliquot of the Reactor 3 effluent was filtered prior to toxicity testing during Rounds 3 and 4. The objective of this testing was to determine if filterable materials were contributing to the toxicity of Reactor 3 effluent and to determine if toxicity would decrease with enhanced secondary clarification. As shown in Figure 7, toxicity of Reactor 3 effluent was reduced substantially by filtration. Although the solids removal achieved by filtration may be greater than that achieved in a full-scale system, the results for the filtered sample are believed to be more representative of the toxicity that would be expected for a system, such as the bioplant, that uses polymer to improve secondary clarification. It should be noted that polymer was not used in this study due to the configuration of the bench-scale reactors.

The LC50 values obtained for the filtered Reactor 3 samples were similar to those obtained for Reactor 1 during Rounds 1 and 2. In addition, the values were similar to those typically observed for the bioplant effluent. These data indicate that more effective removal of toxicity was achieved by treating the problematic waste stream separately with an acclimated activated sludge culture, and filtering the effluent with a pressure filter.

3.2.4 Combined Reactor 2 and Reactor 3. One of the study objectives was to evaluate the feasibility of achieving overall improvement in the biological removal of toxicity under an alternative treatment strategy at the bioplant. The results for Reactor 2 and Reactor 3 suggest that improved COD conversion and toxicity removal was in fact achieved by this strategy. The toxicity of the combined Reactor 2 and Reactor 3 effluents can be predicted by using TUs and flow contribution data to perform a toxicity balance calculation.

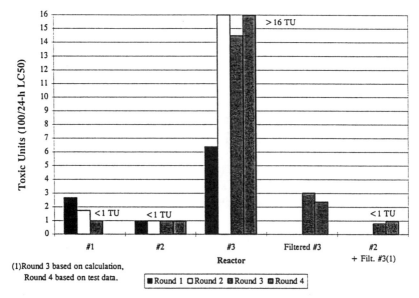

Figure 7 *24 Hour Mysid Data for Reactor Effluent Samples*

As shown in Figure 7, based on the 24 h results from Round 3, the calculated toxicity of the combined Reactor 2 and filtered Reactor 3 samples would be 0.8 TUs (LC50 (>)100% effluent, not acutely toxic). This calculation assumes that the Reactor 2 effluent would contribute 0 TUs and that the filtered Reactor 3 effluent would contribute 0.8 TUs (3.0 TUs x flow contribution to the bioplant of 28% = 0.8 TUs). A combined sample of Reactor 2 and filtered Reactor 3 effluent was tested during Round 4 to confirm that additive and/or synergistic effects would not result in a more toxic sample after combining the waste streams. As shown in Figure 7, the 24 h LC50 for the combined sample was (>)100% effluent (<1.0 TU, not acutely toxic). These data agree with the toxicity predicted by calculation for Round 4 (0.7 TUs, LC50 (>)100%). The Round 4 48 h results for the combined sample did indicate some acute toxicity (1.8 TUs, LC50 = 56.1% effluent); however, the sample was less toxic than the Reactor 1 effluent during Rounds 1, 2, and 3. In addition, the sample was substantially less toxic than the typical bioplant effluent. The predicted 48 h TUs for this sample (1.9 TUs) also agree with the measured value (1.8 TUs). These data suggest that the effluent from the alternative treatment strategy would comply with the toxicity requirements specified in the NPDES permit.

4 DISCUSSION AND CONCLUSIONS

The results of this toxicity treatability study provided important information about the sources of effluent toxicity and possible cost-effective toxicity control options for the facility. The study findings confirmed that one major process waste stream, which contributes approximately 28% of the total process wastewater flow, is the primary source of effluent toxicity. This information allowed TRE efforts to be focused on developing toxicity controls for a relatively low-flow waste stream. Possible cost-effective controls for this waste stream included the alternative biological treatment strategy evaluated in this study as well as a variety of pre-treatment controls and manufacturing process modifications that are also under investigation.

This study also showed that it is possible to achieve improved COD conversion and toxicity removal at the bioplant by implementing an alternative biological treatment strategy. The results indicated that COD conversion and toxicity removal were improved when the problematic waste stream was treated separately from the other major process waste streams. In the existing system, where all waste streams are combined and treated together, the activated sludge culture selectively uses compounds that are easily degraded as the primary carbon source. As a result, the more refractory compounds that cause toxicity pass through the system untreated.

By segregating the problematic waste stream, refractory compounds were concentrated in a single reactor that also contained fewer easily degraded compounds. Therefore, the biomass was forced to become acclimated to the refractory compounds as its primary carbon source and toxicity removal was improved. At the same time, the easily degraded compounds were concentrated in a single reactor which also afforded more complete removal in the absence of the problematic wastewater. The net result, when the two effluents were recombined, was that both COD and toxicity were lower and met discharge permit limits.

The alternative biological treatment strategy evaluated in this study appears to be an effective means for reducing effluent toxicity to acceptable levels. Given the flexibility offered by the bioplant configuration (i.e. four aeration basins and two clarifiers) it may be possible to implement the alternative biological treatment strategy (Figure 2) at the existing bioplant with minimal capital expenditure. On the basis of the annualised cost, implementing separate biological treatment of the problem stream (with filtration) was estimated to be nearly an order of magnitude more cost-effective than the GAC slip stream treatment, which has also been shown to remove toxicity effectively ($84,000 per year versus $611,000 per year).

Toxicity Testing as a Practical Tool for Monitoring Industrial Effluents – A Case Study

D. Watson, P. Fawcett and N. Gudgeon

BNFL International Group, Fuel Division, Springfields, Preston PR4 0XJ, UK

ABSTRACT

A two phase study was carried out on effluents generated by BNFL Fuel at Springfields site. Phase 1 of the monitoring programme was carried out in the Autumn of 1992 and comprised Microtox^{TM} assay on 12 effluent streams. These were representative of the site's liquid waste arisings, the final mixed trade effluent, the final effluent after dilution with stormwater, and a sample from the River Ribble (the receiving water) downstream of the discharge point. Based on the results of Phase 1, a second monitoring programme was undertaken which, in addition to Microtox^{TM} assay, used some of the more established toxicity tests i.e. oyster embryo bioassay, juvenile turbot 96 hour LC50 tests, and brown shrimp 96 hour LC50 tests. The results indicated that Microtox^{TM} could be used as a screening tool for assessing toxicity, and would be useful as a tool in implementing a toxic waste reduction exercise. However, although extensive chemical analysis was also carried out on the samples tested no conclusions could be drawn as to which were the most toxic components. It may therefore be necessary to carry out further work to identify the most toxic elements of a particular effluent if implementing a toxicity reduction programme.

1 INTRODUCTION

Traditionally, the release of potentially polluting liquid effluents to surface waters has been controlled by means of chemical-specific discharge consents. From an industrial point of view this approach has led to simple, clear consent conditions against which compliance can be assessed by means of chemical analysis. However, it is recognised that for complex effluents, and/or those containing a wide range of organic compounds, the approach has a number of weaknesses.

From a regulatory point of view, the application of Environmental Quality

Standards, based on single-substance toxicity data, do not necessarily provide a comprehensive safeguard for the environment *per se*. For this reason, there has been growing interest in the use of direct toxicity based consents to enable the net toxic effects of effluents to be assessed and controlled.

In 1992, BNFL published a formal Statement of Environmental Policy which contained a clear commitment to look beyond the conditions of consents and authorisations to ensure that environmental standards were not compromised by Company activities. With this in mind, British Nuclear Fuels Ltd (BNFL) Fuel Division undertook a pilot study to evaluate the toxicity of selected liquid effluents from the Springfields site. This study was carried out in conjunction with Acer Environmental and was considered to be an adjunct to the ongoing programme of chemical characterisation which has culminated in application for authorisation under the Integrated Pollution Control (IPC) regime as a non-ferrous metal process.

2 SCOPY AND OBJECTIVES OF THE STUDY

The study was undertaken in two phases with the second phase being dependent on the outcome of Phase 1. The scope and objectives of each phase are outlined separately within this section prior to a discussion of the detailed results.

Phase 1

This phase was solely concerned with the monitoring of a range of effluent streams using the Microtox^{TM} test system. Microtox^{TM} was selected in view of its recognised value as a rapid screening bioassay and its known interest to UK regulators. Specific objectives included:

- Monitoring, for a trial period, specific effluent streams to assess their toxicity and the degree of variation of toxicity over a one month period. These effluent streams included a wide range of process effluents, the combined site trade effluent, and the final effluent discharged via the Ribble pipeline (i.e. the site trade effluent combined with stormwater). Additionally, river water samples taken downstream of the discharge point were tested.
- Identification of those effluent streams which could benefit from ongoing toxicity monitoring and/or detailed chemical analysis.
- An overall assessment of the benefits and potential limitations of Microtox monitoring at Springfields.

Phase 2

Following the completion of the Phase 1 study it was decided that the test results were sufficiently worthwhile to justify the initiation of Phase 2. This phase aimed to confirm the validity of Microtox™ as an appropriate monitoring system and to calibrate the system against more established toxicity testing techniques. Specific objectives included the testing of a small number of samples using both the Microtox™ system and three well established toxicity testing techniques, namely oyster embryo bioassay (24 hours), juvenile turbot 96 hour LC50 tests and brown shrimp 96 hour LC50 tests. Additionally, it was intended to generate sufficient Microtox™ data to allow 'toxicity limits' to be set.

Comprehensive chemical analysis of the test streams was also undertaken with a view to identifying specific toxic components.

3 RESULTS OF TOXICITY MONITORING PILOT STUDY

Prior to discussing the results in detail, it is useful to outline the effluent arisings from the site. These are illustrated below (Figure 1) by means of a

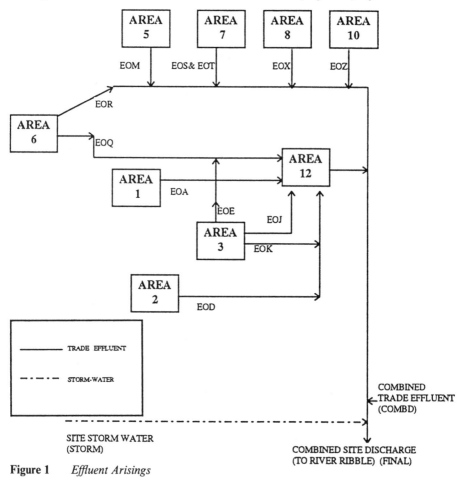

Figure 1   *Effluent Arisings*

schematic diagram which identifies those effluent streams analysed in the study, how they link together to form the combined trade effluent, and where the stormwater joins the trade effluent. Individual streams are denoted by a series of letters; these are sample codes.

Phase 1 Results

Phase 1 of the Microtox™ pilot study indicated that almost all of the individual effluent streams tested were toxic to varying degrees, including the combined trade effluent liquor. However, the final effluent discharge, the site stormwater, and the river water samples taken downstream of the discharge point were either of low toxicity or non-toxic. These results show the importance of mixing and dilution prior to and after discharge.

Detailed analysis of the results from this phase demonstrated that the toxicity of most individual effluent streams was relatively consistent from week to week; full details can be seen in Table 1 below. However, it must be recognised that some toxicity tests do not always produce precise and reproducible results as would be expected from some chemical analysis.

Also of interest was the fact that the majority of samples tended to exhibit slightly higher toxicity over a longer exposure period. This is common to many chemicals and indicates that the toxicity is only manifested after a prolonged period of exposure. However, the 15 minute toxicity results rarely exceeded a two fold difference compared with the corresponding 5 minute

Table 1 *Microtox EC50 Results for Various Streams*

STREAM IDENTITY	5 MINUTES		15 MINUTES	
	MEAN	RANGE	MEAN	RANGE
EOA	14.1	5.1–27.0	13.9	5.8–28.7
EOD	8.7	4.0–12.6	7.5	3.9–10.3
EOE	68.2	47.6–>50	56.9	31.6–>50
EOJ	1.1	0.3–2.3	1.1	0.3–2.3
EOK	36.8	19.3–>50	34.5	16.7–>50
EOM	18.8	1.1–34.8	17.4	1.1–29.4
EOQ	41.4	37.4–47.0	33.3	23.0–45.0
EOR	>50	all >50	68.5	49.0–>50
EOS	19.0	7.5–41.2	10.4	5.6–17.8
EOT	7.9	6.5–9.3	6.2	5.3–6.9
EOX	51.4	17.4–>50	35.2	7.5–>50
EOZ	11.9	8.7–15.8	7.5	4.9–12.1
COMB'D	35.5	10.4–>50	34.8	11.1–>50
STORM	>50	all >50	>50	all >50
FINAL	>50	all >50	68.1	47.5–>50

Notes:
i) " >50" means EC50 greater than 50% (vol/vol)
ii) If all values > 50 then " >50" recorded above; if 1 or more, but not all values > 50 then mean calculated assuming such values as 75%.

Table 2 *Comparison of Microtox^{TM}, Oyster Embryo, Shrimp and Turbot Data*

WEEK	MICROTOX^{TM}		OYSTER 24h EC50	SHRIMP 96h LC50	TURBOT 96h LC50
	5 MIN EC50	15 MIN EC50			
1	22.5–45.0	46*	2.1	7.2	10–18
4	50*	47*	2.6	NT	NT
6	>	>	25.0	NT	NT

Notes:
 i) > means EC50 greater than 45% vol/vol
 ii) NT means not tested
 iii) * means extrapolated

results. Finally, the results from Phase 1 indicated that it may be appropriate to use Microtox^{TM} results to derive toxicity limits for eleven of the effluent streams.

Phase 2 Results

In order to assess if the Microtox^{TM} results were predictive of potential environmental impacts, a sample of the combined trade effluent was tested in a range of standard techniques considered both reliable and environmentally relevant: 24 hour oyster embryo bioassay; 96 hour brown shrimp LC50 test; and the 96 hour juvenile turbot LC50 test. Full details of the results obtained are given in Table 2 below. The oyster embryo bioassay proved to be the most sensitive of the test systems. The ranking in order of sensitivity was oyster embryo bioassay > brown shrimp > juvenile turbot > Microtox^{TM}.

Based on these results, the oyster embryo bioassay was selected for further comparison (calibration) with Microtox^{TM}. Two additional combined trade effluent samples and samples from three process effluent streams were tested using both Microtox and the oyster embryo bioassay. In general, the two tests ranked samples in a similar order in terms of toxicity showing that Microtox results, although less sensitive, were predictive of the results which would be obtained using more established toxicity testing methods.

Additionally, the final effluent and a further nine individual streams were tested using Microtox and the results compiled with the Phase 1 results to give, where possible, a mean EC50 value, a 90 percentile value (i.e. 90% of the samples tested gave results at or above the numerical EC50 value), and an 80 percentile value. The results are presented in Table 3 and indicated that toxicity limits could be used as a means of monitoring effluent quality for a number of streams on site.

Table 3 Percentiles, Ranges & Mean Microtox™ EC50 Results for Various Streams

EFFLUENT STREAM	NUMBER OF SAMPLES	MEAN EC50 (% v/v)	RANGE EC50 (% v/v)	5 MINUTES 80 PERCENTILE	5 MINUTES 90 PERCENTILE	MEAN EC50 (% v/v)	RANGE EC50 (% v/v)	15 MINUTES 80 PERCENTILE	15 MINUTES 90 PERCENTILE
EOA	10	17.8	5.1 - 42.9	8.4 - 10.2	5.1 - 8.4	15.6	5.8 - 45.0	7.6 - 8.3	5.8 - 7.6
EOD	10	4.0	0.23 - 12.6	0.35 - 0.78	0.23 - 0.35	3.6	0.25 - 10.3	0.39 - 0.90	0.25 - 0.45
EOE	4	68.2	47.6 - >	*	*	56.9	31.6 - >	*	*
EOJ	3	1.1	0.3 - 2.3	*	*	1.1	0.3 - 2.3	*	*
EOK	9	31.4	.2 - >	7.8 - 10.1	*	25.0	2.2 - >	7.7 - 9.4	*
EOM	3	18.8	1.1 - 34.8	*	*	17.4	1.1 - 29.4	*	*
EOQ	10	32.9	5.6 - >	13.2 - 15.0	5.6 - 13.2	29.7	5.4 - >	11.8 - 13.3	5.4 - 11.8
EOR	4	>	> - >	*	*	68.5	49.0 - >	*	*
EOS	10	23.4	7.5 - >	10.9 - 12.4	7.5 - 10.9	19.5	5.6 - >	7.5 - 9.3	5.6 - 7.5
EOT	10	6.9	3.3 - 13.1	3.9 - 4.2	3.3 - 3.9	4.8	1.7 - 8.8	3.1 - 3.7	1.7 - 3.1
EOX	5	56.1	17.4 - >	17.4 - 38.1	*	43.2	7.5 - >	7.5 - 18.9	*
EOZ	10	20.7	1.0 - >	1.1 - 2.2	1.0 - 1.1	19.0	0.56 - >	1.2 - 4.2	0.56 - 1.2
COMB'D	10	34.0	2.8 - >	8.8 - 10.4	2.8 - 8.8	34.3	2.8 - >	10.0 - 11.1	2.8 - 10.0

Notes:
i) > means EC50 greater than 45% vol / vol
* means insufficient data

Table 4 Chemical Analysis of Various Streams

STREAM	HEAVY METALS ($\mu g\ ml^{-1}$)										OTHERS		MICROTOX™	
	Cd	V	Cr	Fe	Ni	Cu	Zn	As	Pb	Hg	NH_3	F^-	5 min	15 min
EOA1	0.145	5	4.6	750	2.2	5	8.5	6	2.2	0.034	4.2	105	27.0	28.7
EOA2	0.170	4	4.2	1115	5	12	9.5	1.4	5.8	0.042	6	150	5.1	5.8
EOT3	0.460	0.5	13	1000	17.5	18.5	50	0.29	21	0.006	15000	22	6.5	5.3
EOM2	< 0.03	< 0.04	0.07	2.5	0.12	0.6	2	< 0.1	0.04	0.02	1.3	6000	34.8	29.4
EOM4	< 0.03	< 0.04	0.18	< 1	0.06	1.6	1.1	< 0.1	0.02	< 0.02	1.1	18000	1.1 - 2.3	1.1 - 2.3
EOJ1	0.003	0.006	0.022	3.5	22	9	0.06	0.94	0.007	0.03	1.8	21000	0.6 - 1.1	0.6 - 1.1
EOS3	0.02	0.0035	0.014	0.75	1.9	1.3	3.6	0.09	0.01	0.034	7700	114	7.5	7.5
EOS4	0.0055	0.0016	0.082	2.5	0.48	0.38	2.8	0.003	0.012	0.0018	3000	320	41.2	17.8
EOX1	0.015	1.9	215	1100	168	225	1.3	270	6	0.0055	0.8	580	> 50	39.4
EOX4	0.019	0.6	73	302	34	1.28	0.72	0.2	0.14	0.034	23	310	17.4	7.5

4 CHEMICAL ANALYSIS

Detailed chemical analysis was also carried out on the effluent streams tested (see Table 4) and stepwise regression analysis was applied to attempt to identify toxic constituents.

Although it did not prove possible to establish with certainty which toxicants were responsible for the observed effects, this analysis did give some indication as to the most likely species, in particular Ni ,Cu, Cr, NH_3 and F^-. Further laboratory-based testing is underway with the long term aim of targeting particular discharges for toxicity reduction.

5 CONCLUSIONS

Microtox^{TM} is a low cost and rapid toxicity testing system which could be used to monitor a number of Springfields effluents. It has been shown to be predictive of the results of more established toxicity testing techniques, with each test ranking the toxicity of samples in a similar order. Additionally, the oyster embryo bioassay was shown to be in general 10 to 20 times more sensitive than Microtox^{TM}.

Notwithstanding, Microtox^{TM} does have a number of inherent limitations which have implications for the implementation of an ongoing monitoring programme. These include the necessity to adjust all samples to a narrow pH range as part of the analytical protocol. This was of particular relevance in the case of Springfields effluents which contain a large proportion of potential toxicants in an insoluble form; despite liming, the final effluent is highly variable in terms of pH.

Anomolous results were observed during the Phase 2 work where there was an unavoidable delay between sampling and testing; this clearly emphasised the importance of *in situ* testing.

Microtox^{TM} did prove potentially valuable as one means of identifying "significant effects" as defined in the British Standard BS7750. However, the results of the testing programme were not sufficient, in themselves, to define a toxicity reduction programme, as more detailed testing was necessary to identify specific toxicants.

Since completing the Microtox^{TM} work, and recognising some of the possible limitations, a benthic survey of the Ribble Estuary was carried out in an attempt to assess the potential chronic effects of Springfields discharges. Unfortunately, the dynamics of the estuary were such that no conclusive results were obtained. Solid phase Microtox^{TM} testing of sediment samples, however, did show evidence of toxicity at specific locations with fine-grained sediments.

Toxicity Reduction Measures for a Synthetic Fibre Industry Effluent – A Case Study

J. L. Musterman,[1] T. H. Flippin[2] and G. W. Pulliam[2]

[1]Eckenfelder Black and Veatch, Ltd., Grosvenor House, 69 London Road, Redhill, Surrey RH1 1LQ, UK
[2]Eckenfelder Inc., 227 French Landing Drive, Nashville, Tennessee 37228, USA

ABSTRACT

A series of batch treatability screening tests was conducted to identify technologies suitable for removal of ethylene diamine (EDA) and aquatic toxicity from a synthetic fibre industry works effluent discharged to a municipal secondary treatment works. Treatments by air stripping, cation exchange resin, activated silica, macro-reticular resin, granular activated carbon, and bio-hydrolysis were evaluated in the screening tests. Only cation exchange resin and bio-hydrolysis reduced effluent toxicity to the extent required by the municipal treatment works. Continuous flow, bench-scale activated sludge treatability tests were conducted over a four month period under simulated warm and cold weather operating conditions. The results confirmed that activated sludge treatment alone could consistently provide greater than 95% BOD reduction and complete EDA hydrolysis, nitrification and toxicity reduction. Toxicity reduction could be accomplished at organic loadings greater than 1 kg BOD/kg MLVSS/day, but EDA hydrolysis caused the effluent NH_3-N concentration to be greater than 300 mg/litre. Therefore, a biological pre-treatment system design for the synthetic fibre works was developed based on an organic loading less than the 0.3 kg BOD/kg MLVSS/day in order to achieve compliance with the regulatory authority's requirement of 25 mg l^{-1} or less NH_3-N. The pre-treatment works have been implemented with successful results for toxicity reduction and NH_3-N compliance, based on several years performance data.

1 BACKGROUND

Since the introduction of nylon fibre to the world market in 1940, the demand for synthetic fibres has steadily increased. Process wastewaters from spandex fibre production contain high concentrations of toluene and ethylenediamine (EDA). The discharge toluene concentration is regulated to less than 0.028 mg l^{-1} by the US EPA's Organic Chemicals, Plastics and Synthetic Fibres (OCPSF) effluent guidelines. Process air emissions and wastewater are treated using granular activated carbon (GAC) adsorption, distillation, and air stripping with toluene recovery as illustrated in Figure 1. These systems are highly effective for control of the toluene discharge, but result in poor removal of EDA. A review of the literature (Verschueren, 1983) regarding the environmental effects of EDA is summarised in Table 1. These data indicate that EDA is moderately biodegradable and highly toxic to certain aquatic organisms.

An existing synthetic fibre mill treats its process air emissions and wastewaters using the scheme illustrated in Figure 1. The liquid effluent flow rate of 190 m^3 d^{-1} is discharged to a local treatment works operated by the Sewage Authority. The sewage works has a capacity of 38,000 m^3 d^{-1} and has experienced chronic non-compliance with its discharge licence for acute toxicity. The licence requires that the effluent causes "less than 20 percent mortality to *Ceriodaphnia dubia (C. dubia)* at a 78 percent dilution." A survey by the Authority indicated that the wastewater discharge from the synthetic fibre works was a major source of aquatic toxicity. In addition, the fibre mill's discharge was also in violation of the Biochemical Oxygen Demand (BOD) and ammonia (NH_3) limits of its discharge licence.

This paper presents the approach and results of a wastewater treatability study that was conducted to reduce the aquatic toxicity, ammonia, and BOD of the discharge from the fibre. The study focused on removal of EDA as the

Table 1 *Summary of Environmental Effects of EDA*

Parameter	Value
Toxicity Threshold (cell multiplication inhibition test), mg l^{-1}	
• bacteria (*Pseudomonas putida*)	0.85
• algae (*Microcystis aeruginosa*)	0.08
• green algae (*Scenedesmus quadricauda*)	0.85
• protozoa (*Entosiphon sulcatum*)	1.8
(*Uronema parduczi Chatton-Lwoff*)	52.0
Daphnia LC0, mg l^{-1}	8
Daphnia magna LC50 mg l^{-1}	0.88
BOD_5/Theoretical Oxygen Demand, %	48

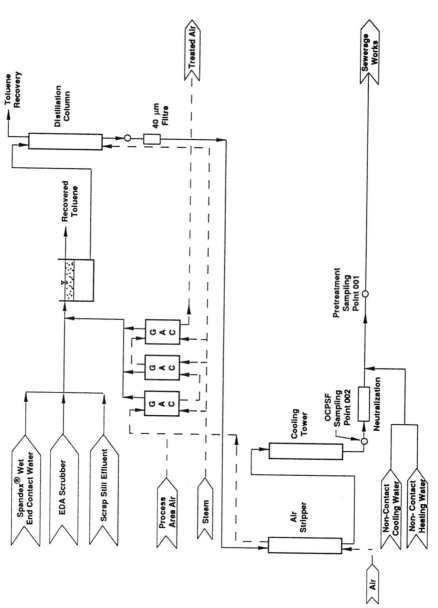

Figure 1 Process Flow Diagram for Treatment of Air and Wastewater

principal toxicant. It consisted of screening level bench-scale treatability studies of selected unit processes to evaluate their toxicity removal performance. These were followed by extended continuous flow studies of a proposed treatment system to demonstrate process performance and develop design criteria. The study results were used to design and construct a pre-treatment facility for the mill. The facility has been in operation for three years with full compliance of its discharge licence.

2 STUDY METHODS AND MATERIALS

Initially, batch type treatability tests were conducted (Phase I) to identify the most effective treatment process(es) for subsequent extended treatability testing (Phase II). Based on the premise that the effluent toxicity was due to EDA, the following technologies were selected for Phase I testing: activated silica, cation exchange resin, macroreticular resin, and aerobic biological treatment. Granular activated carbon and air stripping were also evaluated to broaden the range of potential toxicant removal. Wastewater samples from the fibre mill were composited over a 24 hour period at Outfall 001 (Figure 1) for use in the treatability studies.

Based on the Phase I tests, activated sludge was selected for Phase II testing. Two continuous flow, complete mix activated sludge units were operated to simulate cold and warm weather operating conditions (10 °C to 20 °C) at food-to-microorganism (F/M) ratios of 0.10 to 0.20 kg BOD/kg MLVSS/day. (BOD is 'biochemical oxygen demand', typically measured over five days, which is the commonly used approximation of the 'food' for biological treatment organisms. 'Chemical oxygen demand' (COD) or other parameters can also be used. MLVSS is 'mixed liquor volatile suspended solids' which is the commonly used approximation of microbial mass in biological treatment systems). The units were seeded with a non-nitrifying activated sludge obtained from the sewage works that received the fibre mill's discharge. Approximately six weeks following start-up, both units began nitrifying and data collection began. Prior to the onset of nitrification, effluent from the activated sludge reactors was subjected to high pH batch air stripping tests to determine if this technology (rather than biological nitrification) could be used cost-effectively for NH_3 removal.

The toxicity testing requirements for the fibre mill discharge are illustrated in Figure 2. The first step determined if the mill's toxicity limit (20 percent mortality at 1.78 percent dilution) would be exceeded at the maximum untreated flow contribution from the mill. If the limit was exceeded, then a Batch Activated Sludge (BAS) test was conducted to simulate treatment of the mill discharge at the sewage works. If the BAS treated mill discharge still exceeded the discharge licence of the sewage works (20 percent mortality at 78 percent dilution), then the fibre mill discharge was considered toxic and pre-treatment would be required.

3 RESULTS AND DISCUSSION

The results of the Phase I batch treatment studies and Phase II continuous flow studies are presented in the following subsections. The fibre mill's discharge characteristics and the pre-treatment limits are summarised in Table 2. The discharge was in chronic non-compliance for BOD and acute toxicity. Furthermore, the pre-treatment limit for NH_3-N (25 mg l^{-1}) would be exceeded due to bio-hydrolysis of the influent total kjeldahl nitrogen (TKN) if an aerobic biological process was selected for pre-treatment unless an ammonia removal process was provided.

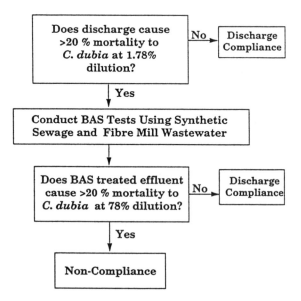

Figure 2   *Toxicity Test Requirement for Fibre Mill Discharge to Sewage Authority*

3.1 Phase I – Batch Treatability Screening Tests

The Phase I treatability study logic diagram is presented in Figure 3 and the results are summarised below.

Air stripping and GAC (Calgon Carbon, Filtrasorb 400) treatment of the mill effluent did not provide significant carbonaceous biochemical oxygen demand (CBOD) reduction. Toxicity analyses of the influent and GAC treated effluent samples revealed no reduction.

Macroreticular resin (Rohm and Haas, Amberlite XAD-4) treatment at a 6 g l^{-1} dosage was investigated on two samples of the mill effluent. CBOD reductions of 93 percent were observed but the resin exhaustion rate was greater than 10 kg kg^{-1} BOD removed. Due to the high capital cost of the resin and the regeneration requirements, further investigations were not conducted.

Batch tests were conducted using weak acid (Rohm and Haas, IRC 50) and

Table 2 *Fibre Mill's Discharge Characteristics and Pre-treatment Limits*

Parameter	Discharge Characteristics [a]	Daily Maximum
Flow, m^3 d^{-1}	190 to 605	675
TBOD, mg l^{-1}	470 to 1,350	250
TCOD, mg l^{-1}	600 to 1,500	2,500
TSS, mg l^{-1}	5 to 45	200
TKN, mg l^{-1}	15 to 300	No Limit
NH_3-N, mg l^{-1}	15 to 60	25
pH	8 to 8.5	6.0 to 9.0
C. dubia Toxicity, % mortality at 1.78% dilution	30 to 100	20
EDA, mg l^{-1}	15 to 1,200	No Limit

[a] The total plant discharge included effluent sources besides the spandex production line.

strong acid (Rohm and Haas, IR 120) cation exchange resins and activated silica gel. These processes provided adequate CBOD and toxicity reduction but required a resin dose in excess of 10 g l^{-1}. Due to the cost of the resin and the volume and problems of regenerant disposal, no further investigation of resin treatment was conducted.

Batch activated sludge treatment at floc loading rates of approximately 0.20 kg BOD/kg MLVSS/day consistently achieved toxicity reduction despite elevated concentrations of effluent CBOD. The results indicated that toxicity reduction by aerobic biological treatment was independent of CBOD reduction. This finding further indicated EDA as the primary toxicant of concern since bio-hydrolysis of EDA to ammonia would occur independently of BOD reduction. The activated sludge, however, was inhibited as evidenced by poor CBOD reduction at conventional floc loading rates and the improved CBOD reduction exhibited under prolonged aeration time. Based on these results, it was concluded that, with proper acclimation, activated sludge treatment would provide reliable pre-treatment compliance and the process was recommended for subsequent testing in Phase II.

3.2 Phase II - Continuous Flow Tests

Pre-treatment for BOD and toxicity reduction could consist of either high rate activated sludge (F/M approximately 1.0/day) followed by ammonia stripping or low rate activated sludge (F/M approximately 0.1/day) with biological nitrification. The results of low rate (mean F/M approximately 0.09/day) activated sludge treatment under simulated cold weather (10°C) operating conditions are presented in Figures 4 to 8.

Figure 4 indicates that the treated effluent soluble BOD (SBOD) and soluble COD (SCOD) concentrations during the study period were consistently less than 30 mg l^{-1} and 250 mg l^{-1}, respectively. The observed "particulate" contributions of BOD and COD in the treated effluent were 0.3 kg BOD and 1.4 kg COD per kg of effluent total suspended solids (TSS). Therefore, if the

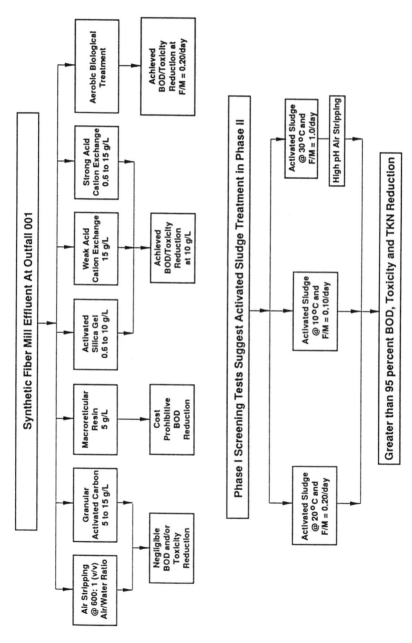

Figure 3   *Treatability Study Logic Diagram*

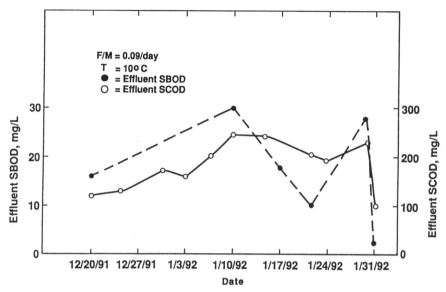

Figure 4 *Chronology of Effluent SBOD and SCOD Concentrations*

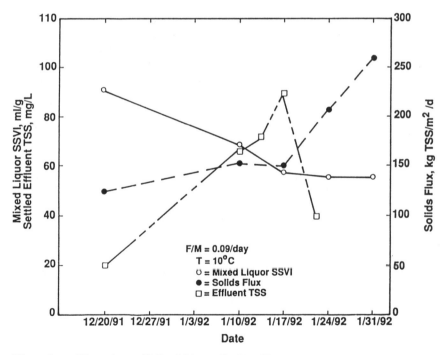

Figure 5 *Chronology of Mixed Liquor Settling Characteristics*

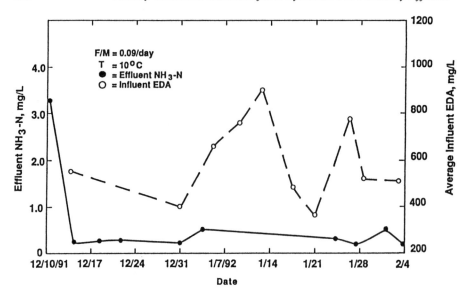

Figure 6 *Chronology of Influent EDA Loading and Effluent NH₃-N*

treated effluent TSS concentration were maintained at the licence limit (200 mg l⁻¹), the resulting effluent BOD and COD concentrations would be approximately 90 mg l⁻¹ and 530 mg l⁻¹, respectively. The BOD and COD discharge limits are 250 mg l⁻¹ and 2,500 mg l⁻¹, respectively. Figure 5 presents a chronology of the reactor effluent TSS concentrations and stirred sludge volume indicator (SSVI) and solids flux data for the mixed liquor during the study period. The results indicate that the mixed liquor settled well (SSVI = 58 to 91 ml g⁻¹) and produced settled supernatant TSS concentrations of 12 to 90 mg l⁻¹.

Figures 6 and 7 present chronologies of reactor influent EDA and effluent NH₃-N concentrations (Figure 6) and effluent toxicity (Figure 7) during the study period. The effluent NH₃-N concentrations were consistently less than 1.0 mg l⁻¹ after a stable nitrifying biomass was established and adequate alkalinity was provided. During this period, the influent EDA concentrations varied from approximately 300 to 900 mg l⁻¹ and there was no detectable EDA in the reactor effluent. The effluent *C. dubia* toxicity, at 1.78 percent dilution, was essentially zero throughout the study period (Figure 7). Based on these results, it was concluded that the low rate activated sludge process with nitrification would satisfy the pre-treatment limits for BOD, TSS, NH₃-N and toxicity.

Figure 8 presents the results of mathematical modelling for process design of an activated sludge plant to provide pre-treatment of the fibre mill wastewater. The modelling results indicated that compliance with the ammonia pre-treatment limit of 25 mg l⁻¹ (rather than the BOD limit) dictated the process design conditions for F/M. Effluent BOD compliance could be achieved at an operating F/M greater than 3/day, whereas an operating F/M less than 0.3/day

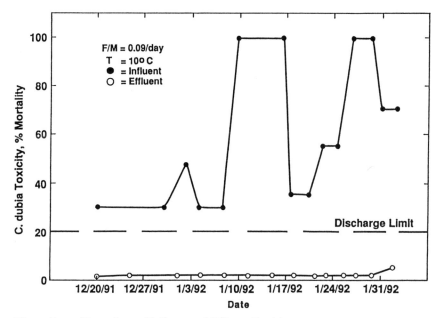

Figure 7 *Chronology of Influent and Effluent Toxicity*

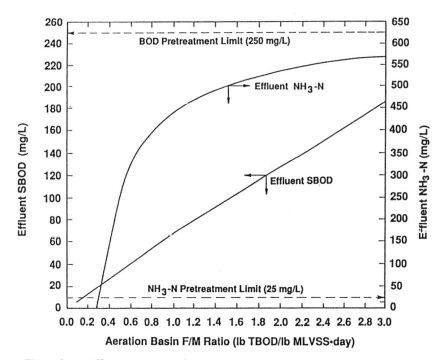

Figure 8 *Effluent NH_3-N and SBOD Concentration as a Function of Aeration Basin F/M Ratio at 30 °C*

was required for effluent NH_3-N compliance. Due to the increased aeration basin tankage requirements for a F/M of 0.3/day, however, the high F/M system for BOD removal followed by air stripping for NH_3 removal was further evaluated. Air stripping tests at pH 10 to pH 11 were conducted on the activated sludge effluent prior to the onset of nitrification. The results indicated that an air : water ratio (A:W) of approximately 12,000:1 (v/v) was required for licence compliance. Since local air emission regulations for ammonia would require capture and treatment of the off-gas (approximately 95,000 m^3 h^{-1}), this treatment alternative was abandoned and the low rate activated sludge process with biological nitrification was selected.

4 PROCESS DESIGN OF PRE-TREATMENT FACILITY

The results of the Phase II treatability study were used to develop process design criteria and operational parameters for the full-scale pre-treatment facility. These are summarised in Table 3. A flow schematic of the recommended pre-treatment system is shown in Figure 9.

The total construction cost was approximately US$1,100,000 and the estimated first year annual cost for operation, maintenance, materials, chemicals and power was US$160,000/year. The estimated operations and maintenance costs for the pre-treatment facility were developed based on the following cost data (1994 US dollars):

- 114,000 kwh per year at US$0.033/kwh
- 2,500 labour-hours at US$40/hour
- US$10,100 per year in maintenance materials
- US$46,000 per year in chemical costs

Table 3 *Process Design and Operational Criteria for Pre-treatment Facility*

Parameter	Value
Flow Rate, m^3 d^{-1}	190
BOD/TSS effluent, kg kg^{-1}	0.3
COD/TSS effluent, kg kg^{-1}	1.4
Reaction Rate on BOD/COD basis, day^{-1}	20.8/5.0 @ 30 °C
Heterotrophic Yield, kg MLVSS/kg BOD removed	0.51
Nitrification Rate, kg NH_3-N removed/kg $MLVSS_N$ day^{-1}	0.95 @ 30 °C
Oxygen Required, kg O_2 required/kg BOD removed	0.84
Sludge Age, days	40
MLVSS/MLSS, kg kg^{-1}	0.8
Clarifier Hydraulic Loading Rate, m^3 m^{-2} d^{-1}	10
Clarifier Solids Loading Rate @ 100% Recycle, kg TSS m^{-2} d^{-1}	60–180

* $MLVSS_N$ is the mass of MLVSS estimated to be nitrifying microorganisms.

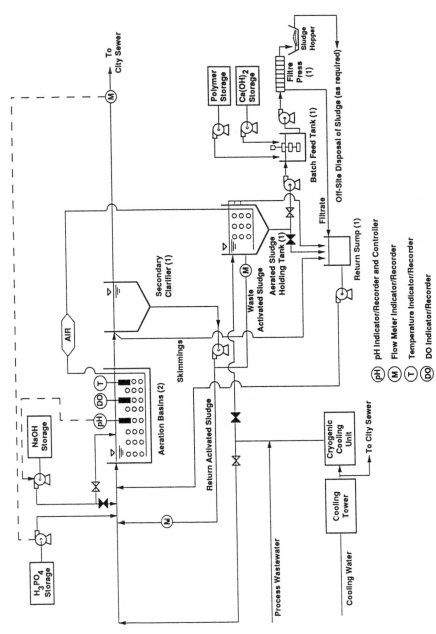

Figure 9 *Process Flow Schematic for Pre-treatment Facility*

5 OPERATIONAL PERFORMANCE

In the year preceding installation of the pre-treatment facility, the fibre mill's monthly average discharge BOD concentration ranged from 109 to 1,256 mg l^{-1} with a mean of 794 mg l^{-1} (licence limit = 250 mg l^{-1}). During this period the discharge consistently exceeded its toxicity limit at a 1.78% dilution. The pretreatment facility was placed in service in June 1994 and by August 1994 full nitrification had been established. The average discharge NH_3-N concentration in August was reduced to 7 mg l^{-1} (from 600 mg l^{-1}) and there was no acute or chronic toxicity to *C. dubia* at 1.78% dilution (the licenced discharge flow rate contribution). Effluent analyses in September 1994 indicated a NH_3-N concentration of 3 mg l^{-1} and no acute toxicity at 1.78% dilution. Furthermore, the discharge exhibited no chronic toxicity to *C. dubia* at a 75% dilution as tested by the Sewage Authority. The average annual discharge BOD and NH_3-N concentrations during 1995 were 21 mg l^{-1} and 5 mg l^{-1}, respectively. The discharge is no longer tested for aquatic toxicity since it has been in full compliance with its licence.

6 CONCLUSIONS

The following conclusions were developed:

1. EDA was the principal toxicant to *C. dubia* in the synthetic fibre mill's effluent. EDA, however, was readily bio-hydrolysed to ammonia and other organic constituents that were biodegraded and nitrified by an acclimated biomass.
2. A completely mixed activated sludge with nitrification operated at a F/M of approximately 0.1/day at 25 °C has provided consistent compliance for nearly two years with pre-treatment limits of 250 mg l^{-1} for BOD, 200 mg l^{-1} for TSS, 25 mg l^{-1} for NH_3-N, and <20 percent mortality for *C. dubia* toxicity at 78% effluent dilution.
3. The completely mixed activated sludge system performance for BOD removal and nitrification has been stable under influent EDA concentrations of 200 to 1,200 mg l^{-1} (average = 550 mg l^{-1}) and BOD concentrations of 530 to 1,130 mg l^{-1} (average = 918 mg l^{-1}).

References

1. J. L. Musterman, and Y. Argaman, (1991). Off-Gas Recycle for Enhanced VOC Biodegradation - An Alternative VOC Control Strategy. Presented at 64th Annual WPCF Conference in Toronto, Ontario.
2. US EPA (1993). Effluent Guidelines and Standards for Organic Chemicals, Plastics, and Synthetic Fibers. 40 *CFR* 414, 58 *FR* 36892.
3. K. Verschueren, (1983). *Handbook of Environmental Data on Organic Chemicals*, Second Edition, Van Nostrand Reinhold, New York.

What Next ? – Future Developments in the Field of Aquatic Environmental Toxicology

I. Johnson

WRc plc, Henley Road, Medmenham, Marlow SL7 2HD, UK

ABSTRACT

The assessment of the impact of wastes on the aquatic environment and subsequent control measures have largely been based on a chemical-specific approach in which the release of individual substances is limited. However, a Direct Toxicity Assessment approach is now being increasingly advocated and used by regulators in the United Kingdom to control effluent discharges and to monitor receiving water quality.

At present, the tests used are mainly those required for product registration, that is whole organism algal, invertebrate and fish tests with lethal and sublethal (growth and reproduction) endpoints. However, the duration and cost of these tests means they are inappropriate for providing real time measurements of toxicity. Consequently, attention has focused on the use of rapid, sub-organism and microbial system assays including those where genetic engineering is being used to employ easily measured endpoints in ecologically relevant species.

This paper considers how these rapid systems will be developed, particularly into continuous monitors, and how they may be used in the future in an integrated regulatory framework with chemical and biological assessment tools. It considers what requirements tests will have to meet in terms of sensitivity, spectrum of response, precision and ecological relevance to be of use in specified roles.

1 INTRODUCTION

In considering what future developments may take place in the field of aquatic environmental toxicology it is initially worth revisiting the purpose of controlling the release of wastes to the aquatic environment which is *to protect* populations and communities in aquatic ecosystems from the adverse effects of synthetic contaminants.

Traditionally, the assessment and control of effluents discharged to the aquatic environment has been carried out by a chemical-specific approach.[1] The potential impact of the discharge on the receiving water community has been evaluated by reviewing available toxicological data for identifiable substances of concern. Control has been achieved by establishing maximum permitted concentrations of the relevant contaminants in the discharge which will satisfy established Environmental Quality Standards (EQSs) or predicted no effect concentrations (PNECs) for substances of concern. Appropriate sampling and chemical analysis of the waste is undertaken to ensure the permitted levels are not exceeded.

The approach has been found to be satisfactory for simple discharges of well defined and consistent composition containing only contaminants for which there are established EQSs or adequate toxicological data to derive appropriate PNECs. The consent conditions derived in such cases are clear cut. In addition, compliance monitoring by chemical analysis of a limited number of determinants is relatively inexpensive and, with appropriate analytical quality control, can give adequate accuracy and comparability.

Monitoring the quality of receiving waters has traditionally involved both chemical analysis of water samples and, in rivers, biological surveys of benthic macro-invertebrate communities.

However, there is now growing recognition that a Direct Toxicity Assessment (DTA) approach can be valuable for assessing and controlling the toxicity of discharges and monitoring the quality of receiving waters. The Environment Agency and the Scottish Environment Protection Agency are currently working on the introduction of toxicity-based criteria for the regulatory control of discharges and the use of DTA both as part of a General Quality Assessment (GQA) Scheme and in local Environmental Impact Assessments.[2]

2 INTEGRATED APPROACH

In coming years it seems certain that an integrated (*triad*) approach using chemical, environmental toxicological and biological assessment tools will be used by regulators to control discharges to aquatic systems and for associated monitoring of the quality of receiving water column and sediment. Each component of the triad has its own advantages and limitations (Table 1) but none used in isolation usually provides sufficient information to make effective regulatory decisions. However, in combination they complement each other and offer a more complete approach to the protection of controlled waters. Experiences from North America indicate that, if applied effectively on a catchment basis, an integrated approach allows existing discharges to be controlled and environmental improvements to be detected and demonstrated.[3] The requirements of Integrated Pollution Control and the need to

Table 1 *Advantages and Limitations of Chemical, Biological and Environmental Toxicological Assessment Tools*

Approach	Advantages	Limitations
Chemical assessment	1 Numeric standards can be set	1 Not all contaminants present are considered
	2 Appropriate for simple, well defined discharges	2 Interaction of contaminants not considered
		3 Numeric standards set may not be protective
Biological assessment	1 All stressors measured	1 Numeric standards are not defined
	2 Ultimate endpoint	2 Cause of impact not identified
Environmental toxicological assessment	1 All contaminants present are considered	1 Incomplete data on causative agent
	2 Interactions are considered	2 Endpoints may not be predictive of ecosystem response
	3 Numeric standard can be applied	

consider the Best Practicable Environmental Option for the release of wastes means that the integrated approach should ultimately be applied to the other environmental compartments of air and land as well as aquatic systems.

3 ROLES IN WHICH ENVIRONMENTAL TOXICOLOGICAL TOOLS ARE NEEDED

The roles in which environmental toxicological tools are primarily needed are:

1. regulation of point discharges (episodic and continuous) to aquatic systems (that is end of pipe assessment); and
2. monitoring of the quality of receiving waters and sediments in response to both point and non-point (diffuse) sources of pollution.

The regulation of discharges includes assessing the toxicity of individual substances to provide data for the derivation of EQSs and PNECs. In controlling the toxicity of final discharges to protect receiving water populations and communities, it may be increasingly necessary to use environmental toxicological tools to identify toxic waste streams in an industrial plant or toxic inputs at a sewage treatment works which result in non-compliance of a final discharge with regulatory control measures.

Monitoring of receiving water quality may involve the use of toxicological tests in general quality assessments, local environmental impact assessments and the investigation of pollution incidents. Consideration of the toxicity of sediments is a key area in future catchment management planning since sediments are now recognised as one of the significant factors which can lead to the degradation of surface water quality and adverse effects on benthic communities and fish populations.[4]

4 TYPES OF ENVIRONMENTAL TOXICOLOGICAL TOOLS NEEDED

For the regulatory roles described in Section 3 two principal types of tools are needed:

1. portable and continuous (on-line) monitors; and
2. a battery of acute and chronic lethal and sub-lethal tests for the water column and sediment.

In these roles, tests are needed which consider not only traditional toxic effects (such as those on growth, reproduction, behaviour and survival) but also genotoxic impacts and the effects of specific acting substances such as oestrogens. The tests could be applied at the sub-organism, organism or population level of biological organisation.

The portable and continuous monitors could be either semi- or fully automated bioassays or be based on biosensor technology. Portable systems will be used to track toxicity within industrial plants and monitor receiving water toxicity within a catchment. Continuous monitors will be applied at the end of discharge pipes and at key monitoring points in a catchment. Ultimately, it is likely that these systems will be linked telemetrically to regulatory facilities.

The battery of acute and chronic tests could be laboratory-based and carried out using samples collected in the field, or could be deployed *in situ*. Single-species or multi-species tests could be used depending on application.

The two types of tests, desirable characteristics, current availability and future developments will be discussed in the following sections.

4.1 Portable and Continuous (On-line) Monitors

4.1.1 Characteristics. Portable and continuous monitors to be used in assessing and controlling the toxicity of discharges released to the aquatic environment (or other compartments) ideally need to:

1. generate data rapidly to allow real time control so that discharges which may breach regulatory control measures can quickly be identified and dealt with (for example by diverting the discharge to holding tanks and releasing it in a controlled fashion which will not impact the receiving water);
2. be cost effective so they can be applied widely;

3 be sensitive to a wide range of contaminants so they respond to any changes in discharge composition such as may occur for batch processes; and
4 be robust and easy to use so they lend themselves to self monitoring and do not suffer from excessive down time (and for continuous monitors, false alarms).

Ideally, the monitors should use environmentally relevant endpoints and ecologically relevant species (that is species which are indicator organisms for an ecosystem or which fulfill a key functional role). However, devices using less relevant species and endpoints can be used providing the response is correlated with environmentally relevant endpoints in ecologically relevant species. These correlations should be more robust and predictive if they can be shown to be mechanistically based (that is reflecting the modes of action of the contaminants involved).

4.1.2 Current availability. A number of commercial, portable and continuous (on-line) toxicity monitors are available measuring endpoints such as bioluminescence and chemiluminescence, oxygen consumption, metabolic activity, fluorescence, motility, shell valve activity and changes in ventilation rate (Table 2[5] and papers presented at this conference, by Upton and Pickens, Bolton *et al*, Brown *et al*, Rogerson, Colley *et al* and Hayes and Smith). In these systems sub-cellular components and micro-organisms (such as bacteria and algae) are generally used, though systems are available using crustaceans, molluscs and fish.

Many of the systems described in Table 2 have only been developed in the last 5 years and have not been widely applied to assessing the toxicity of discharges or receiving waters. Increased application of these devices in coming years, and further consideration of the information they can provide, will result in a clarification of which systems are most appropriate for particular roles.

4.1.3 Future developments. Current developments with genetic engineering offer the possibility of inserting genes for easily measured endpoints (such as light production) into ecologically relevant species as markers for particular processes. For example *lux*-modified cultures of the important rhizobacterium *Pseudomonas fluorescens* have been produced.[6] These genetically modified organisms have been shown to be more sensitive to the contaminants cadmium, copper and zinc than the marine bacterium *Photobacterium phosphoreum (Vibrio fischeri)* used in the Microtox^{TM} test by factors of 21-58 times (Table 3).[7] For dichlorophenol there were no differences in sensitivity between the test species.

The *lux*-modified bacteria have also been used along with *P. phosphoreum* to assess the toxicity of a whisky distillery effluent and the receiving water in the immediate vicinity of the discharge (Table 4).[8] As for the pure substances, the lux-modified bacteria were more sensitive than the Microtox^{TM} bacteria.

Table 2 *Commercially Available Portable and Continuous (On-line) Monitors*

Type of monitor and System	Monitor endpoint	Test duration	Test species
Portable			
ECLOX	Light production (Chemiluminescence)	4 min	Horse radish peroxidase complex
Aquanox	Light production (Chemiluminescence	4 min	Horse radish peroxidase complex
Microtox	Light production (Bioluminescence)	5-30 min	*V. fischeri*
TOXALERT	Light production (Bioluminescence)	5-30 min	*V. fischeri*
POLYTOX	Oxygen consumption	20-30 min	12 strains of bacteria
Continuous			
Microtox-os	Light production (Bioluminescence)	15-20 min	*V. fischeri*
Biox 1000T	Oxygen consumption	No data	Bacterial biofilm
Stiptox norm	Oxygen consumption	No data	*P. putida*
Toxalarm	Oxygen consumption	40 min	*P. putida* or algal spp
Toxiguard	Oxygen consumption	No data	Bacterial biofilm
Eu-Cyano bacterial electrode	Amperometry	< 60 min	*E. coli* and *Synechococcus*
Biosens-algae toximeter	Fluorescence	30 min	*C. reinhardii*
DF-algae test	Fluorescence	No data	*C. reinhardii*
Algentest IWF fluorometer	Fluorescence	20 min	*C. reinhardii*
FluOX test	Fluorescence	< 30 min	*C. reinhardii*
Aqua-Tox Control *Daphnia* monitor (Kerran)	Motility		*D. magna*
Mossel Monitor (Delta Consult)	Shell valve activity	1 min	Bivalves
Fish monitor (ELE)	Change in ventilation rate	30 min	*O. mykiss*
Aqua-Tox Control Fish monitor (Kerran)	Rheotaxis	No data	*L. idus* or other species

(TM Symbols ommitted from this table.)

Table 3 *Comparison of Sensitivity of Lux-modified* P. fluorescens *and* P. phosphoreum *to a Range of Contaminants*

Test species	IC50 (mg l^{-1}) values for selected contaminants			
	Cadmium	Copper	Zinc	Dichlorophenol
P. fluorescens	0.17	0.09	0.09	1.86
P. phosphoreum	9.78	1.89	2.35	1.68

Table 4 *Comparison of Sensitivity of Lux-Modified* P. fluorescens *and* P. phosphoreum *to a Whisky Distillery Effluent and Receiving Water Samples*

Test species	Bioluminescence as a percentage of upstream control		
	Upstream	Effluent	Downstream
P. fluorescens	100	64.0	76.0
P. phosphoreum	100	92.5	95.6

In the future, these rapid systems may also have a diagnostic capability and be able to identify the causative agents responsible for toxicity in samples. This would involve having variants of the test system which had been produced to respond to particular classes of compounds with specific modes of toxic action (for example non-polar narcotics, respiratory blockers and cholinesterase inhibitors). In the variants, genes for easily measured endpoints such as light production would be inserted to activate when changes occurred at specific points in biochemical pathways. In the example in Figure 1, the toxicity of a sample to the non-specific test has been identified as being due to cholinesterase inhibitors using the chemical class specific variants.

A collaborative project by Brunel University, the National Rivers Authority and the Ministry of Agriculture Fisheries and Food has resulted in a genetically engineered yeast based assay system for identifying oestrogenic substances.[9] In the system, the human oestrogen receptor gene is integrated into the main yeast genome and is expressed in a form capable of binding to oestrogen response elements within a hybrid promotor on the expression

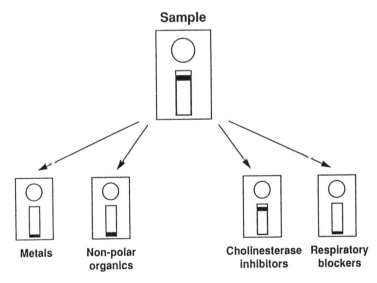

Figure 1 *Use of a Portable Meter to Determine the Toxicity of Environmental Samples and Identify the Causative Agent*

plasmid. Activation of the receptor by binding of ligand, causes expression of the receptor gene *lac*-Z which produces the enzyme β-galactosidase. This enzyme is secreted into the medium and metabolises the chromogenic substrate CPRG from a yellow to red product which can be measured by absorbance.

4.2 Battery of acute and chronic lethal and sub-lethal tests

Laboratory-based and *in situ* tests are needed to:

1. assess the toxicity of discharges and regulate effects at the end of pipe or at designated points in the receiving water; and
2. monitor receiving water quality.

Laboratory-based tests can be used to establish control measures for an effluent to prevent acute and/or chronic effects of the discharge at a specified point in the receiving water. *In situ* tests allow these points to be verified and the success of the control measures to be assessed (see papers presented at this conference by Crane *et al* and Roddie *et al*). *In situ* tests have the advantage of integrating exposure conditions and allowing toxicological assessments to be integrated with:

1. a suite of appropriate biomarkers;
2. measurements of tissue contaminants to assess contaminant bioavailability; and
3. toxicity identification procedures to determine causative agents (see paper presented at this conference by Donkin *et al*).

4.2.1 Characteristics. For the roles described in Section 3, the acute and chronic tests used need to have standard operating procedures, to be repeatable and reproducible, to be sensitive to a wide spectrum of chemicals and to have exposure durations appropriate to the stability of the samples. There are advantages to using the same species for both the acute and chronic tests and for both laboratory-based and *in situ* tests since this allows for the comparison of data using a common toxicological currency.

4.2.2 Current availability. At present there are a number of acute lethal and sub-lethal single species tests at different trophic levels which are available to assess the toxicity of discharges and monitor the quality of the water column and sediments. The tests have standard operating procedures and are recognised by organisations such as the Organisation for Economic Cooperation and Development (OECD), the International Standards Organisation (ISO), the International Commission for the Exploration of the Seas (ICES) and the United Nations (UN). However, there is clearly a need for the development of standardised chronic test methods for testing the toxicity of discharges and receiving water column and sediment samples. The 21-28 day exposure regimes used in current laboratory-based chronic invertebrate and fish water column tests for product registration are not appropriate given the potential instability of environmental samples.

4.2.3 Future developments. Single species acute and chronic tests are likely to remain the key methods for assessing the toxicity of discharges and receiving water column and sediment samples for the foreseeable future. However, in coming years greater use may be made of multi-species tests since they offer endpoints with greater ecological relevance than single species tests.[10] Furthermore, they may be as efficient as an array of individual tests in assessing the toxicity of wastes and may be no more expensive. However, the duration of multi-species tests (typically 21-63 days) may be a problem given the stability of many test samples.

5 CONCLUSIONS

An integrated approach to controlling toxic wastes discharged to aquatic systems is needed combining chemical, environmental toxicological and biological assessment tools.

Environmental toxicological tools are primarily needed for:

1 regulation of point discharges (episodic and continuous) to aquatic systems (that is end of pipe assessment); and
2 monitoring of the quality of receiving waters and sediments in response to both point and non-point (diffuse) sources of pollution.

For regulating discharges and monitoring receiving water column and sediment quality, portable and continuous (on-line) monitors and a battery of acute and chronic higher organism tests are needed. Development and standardisation work is necessary in order to provide the tools for a particular role.

References

1. National Rivers Authority, 'Discharge Consent and Compliance Policy: A Blueprint for the Future.' Water Quality Series No 1. NRA, London, 1990
2. National Rivers Authority, Aquatic Toxicity Control and Assessment: An update on current research and development. NRA, London, 1995
3. P. M. Chapman, *Mar. Poll. Bull.*, 1995, **31**, 167
4. W. J. Adams, R. A. Kimerle and J. W. Barnett, *Env. Sci. Technol*, 1992, **26**, 1865
5. I. Johnson Identification of screening, lethal and sub-lethal toxicity tests for assessing effluent toxicity. NRA R & D Note 389, 1995
6. K. Killham, *Biotechnology*, 1992, **10**, 830
7. G. Paton, C. D. Campbell, L. A. Glover, and K. Killham, *FEMS Lett. App. Microbiol*, 1995, **20**, 52
8. G. Paton, G. Palmer, A Kindness, C. Campbell, L.A. Glover and K Killham, *Chemosphere,* 1995, **5**, 3217
9. E. Routledge, C. Dsbrow, G. Brighty, M. Waldock and J. Sumpter, Identification of oestrogenic chemicals in effluent from sewage treatment works. NRA R & D Note 490, 1995
10. Cairns Jr and D. S. Cherry, In, 'Handbook of Ecotoxicology', edited by P. Calow, Blackwell, Oxford, pp 101-116, 1993

Changes in Toxicity of Ground and Surface Waters during Remediation of a Former Gasworks Site

A. J. Hart and A. Trim

British Gas, Research and Technology, Gas Research Centre, Loughborough LE11 3QU, UK

The use of manufactured gas as a fuel began in the UK in the early 1800s and continued until the advent of natural gas in the mid 1960s. During that time literally thousands of gasworks were constructed and operated by both private companies and local authorities. Although these sites were decommissioned to the environmental standards of the time, some still contain residues from the gasification process with the potential to cause unwanted environmental effects. This poster details the changes in acute toxicity during remediation of a former gasworks site in Yorkshire.

Water samples were collected from boreholes on the site and from adjacent surface water courses (a river and a canal) and Microtox™ analyses carried out to determine acute toxicity. Traditional chemical analyses were also performed. Toxicity measurements were carried out before, during and after major remediation works on the site. In particular, measurements continued for many months after the remediation to ensure that any reductions in toxicity were not merely short term effects.

Initially, significant levels of toxicity were detected at a number of locations on the site. For some of these the toxicity temporarily increased whilst the remediation works proceeded on site. However, once the work was complete and all excavated contamination removed, acute toxicity rapidly fell to below detectable levels and has remained so.

Rapid Determination of Heavy Metal Toxicity in Sewage Sludge Using a Bioluminescence-based Bioassay

S. Sousa,[1] C. D. Duffy,[1] E. A. S. Rattray,[1] K. Killham,[1] L. A. Glover,[2] D. Fearnside,[3] K. C. Thompson,[3] and M. S. Cresser[1]

Departments of [1]Plant and Soil Science and [2]Molecular and Cell Biology, University of Aberdeen, Aberdeen AB9 2UE, UK
[3]Yorkshire Water Services, Western House, Bradford BD6 2LZ, UK

There is concern that the application of sewage sludge to land may have toxic impacts on both the terrestrial and aquatic environment, the latter through leaching of potential pollutants to ground water. Although sewage sludge may contain organic or inorganic pollutants, this poster describes a method for the monitoring of heavy metal toxicity, which currently involves full chemical analysis. This is often too time-consuming for practical management of sludge disposal to land. Even though chemical analysis is required to meet EU legislation, there is the need to develop a rapid, surrogate assay which correlates well with full chemical analysis and plant growth assays. Biological based assays have been used to assess the toxicity of water, with microbial assays providing rapid analysis of material. Bioluminescence-based assays (Microtox and *lux*-based) have been shown to provide the rapidity and sensitivity for assessing the toxicity of waste samples.

Measurement of total metal concentration in sewage sludge generally involves sample digestion to remove the organic matrix and solubilise the metals. For waste preparation to be compatible with a biological assay, total removal of the organic matrix should be obtained in order to release heavy metals completely, ideally with no residue remaining in the final extract. The aim of this study was to determine the compatibility of a novel digestion technique of sewage sludge with a bioluminescence-based microbial bioassay using *lux*-marked *Pseudomonas fluorescens*, to assess the toxicity of heavy metals in sewage sludge.

The total metal concentration was determined in sewage sludge samples from eight different treatment works and digestions prepared to assess the toxicity using two bioluminescence-based bioassays, *lux*-marked *Ps. fluorescens* and *Photobacterium phosphoreum*.

The digestion technique was shown to remove successfully the organic matrix of the sludge and provide a final extract with no residue from the digestion reagent. A series of different solutions was used to extract the heavy metals from the final digest, including EDTA and water at different pH values. The results showed that heavy metal bioavailability was affected by the extraction solution used and the bioluminescence-based bioassays were able to resolve differences in the total metal load of the sludges tested.

Fate and Environmental Effect of Produced Water Discharged from the Clyde Oil Production Platform

I. Vance,[2] E. J. Butler[2] and S. A. Flynn.[1]

[1]BP Group Research and Engineering Centre, Sunbury-on-Thames, Middlesex TW16 7LN, UK
[2]BP Exploration Operating Company Ltd, Farburn Industrial Estate, Dyce, Aberdeen AB1 9TL, UK.

British Petroleum has initiated an extensive study of the composition, toxicity, biodegradability and dispersion characteristics of produced water discharged from the Clyde reservoir in the Central North Sea. Clyde produces approximately 24 mbd oil and 50 mbd water from a single production platform located in 80 m of water, some 300 km South East of Aberdeen.

Compared to data published for other produced water discharges entering the North Sea, Clyde produced water had a relatively low total organic carbon (TOC) concentration of 33 - 45 mg l^{-1}. The TOC was rapidly biodegraded in standard tests, with a loss of 88% after 8 days (mean of three tests). Biodegradation proceeded and reached a value of 97% after 28 days.

Light emission by the bacterium *Photobacterium phosphoreum* (Microtox^{TM} EC50 15 minutes) was more sensitive than the mortality of the copepod *Tisbe battagliai* (EC50, 48 hours) to the toxicants present in Clyde produced water. Loss of volatile organic carbon resulted in a three-fold decrease in toxicity measured by the Microtox^{TM} test. Biodegradation of the remaining organic carbon over 28 days reduced the toxicity by a further factor of two, resulting in a final value of EC50 (15 minutes) of 52% (by volume).

Hydrographic data appropriate to the location of the Clyde discharge were used in a particle tracking mathematical model to predict the dilution of the produced water plume with time and distance from the outfall. Disregarding the effects of biodegradation, a minimum dilution factor of 630 is predicted at a point 100 m from the outfall. At remote locations, 10 km from the outfall, a minimum dilution factor of 15,000 (including the effect of biodegradation) is predicted.

Owing to the relatively low toxicity and high dilution of the discharge, the environmental impact of Clyde produced water is likely to be minimal.

Remediation of Contaminated Dredged Sediments by Physical Processing

S. T. Hall, A. Burton and B. Denby

Institute of Environmental Engineering University of Nottingham, Nottingham, UK

Heavy metals and other toxic pollutants, normally resulting from historical industrial activities, are often present in dredged river, canal and harbour sediments. These pollutants are often contained within a specific particle size and/or density fraction. The contaminants, if present as discrete particles, will often possess different surface properties from the bulk of the dredged sediment particulates. These different physical properties can be utilised to remove the pollutants into a relatively small volume for controlled disposal or further treatment, leaving the majority of the processed material as "clean" sand or silt. However, given the differences in particle size, composition, pollutant type and concentrations in sediment samples from different locations; expert judgement is required in the selection of the appropriate treatment processes and extensive testing is often necessary to assess their level of success.

This poster explains the application of various physical remediation techniques (including size separation, gravity separation and froth flotation) to the treatment of dredged sediments. It presents results from various case studies and describes a prototype knowledge-based ("expert") system that is being developed to assist in the evaluation of such processes in the treatment of contaminated soils and sediments.

Toxicity Assessment of Complex Industrial Effluents Discharged to Sewer

M. Arretxe,[1] G. Ellen,[1] P. Tetlaw,[1] M. Heap[1] and N. Cristofi[2]

[1]East of Scotland Water, Industrial Pollution Control Section, 55 Buckstone Terrace, Edinburgh EH10 6XH, UK
[2]Pollution Research Unit, Napier University, 10 Colinton Road, Edinburgh EH10 5DT, UK

Historically, if industry wished to discharge wastewater into aquatic environments, it would simply provide a list of the constituents contained in the waste. The controlling authority would assess, from available toxicity data, the effects of the constituents on the receiving water and set the consent conditions to provide the required degree of protection (chemical specific approach). Three assumptions are made in this procedure. Firstly, it is assumed that the toxicity of each of the listed substances is known; secondly, that the chemical content of waste is adequately defined and thirdly, that the constituents of the waste do no act antagonistically or synergistically.

The majority of industrial wastewaters do not discharge directly to river. They are discharged to sewer and receive treatment at a wastewater treatment works. Clearly the first concern with such discharges is to ensure that there is no toxic impact on the receiving works. The Public Health Act, 1936, introduced the means of control of potentially toxic substances discharged to sewer. Section 27 prohibited the discharge of matter likely to affect prejudicially the treatment and disposal of sewage. As discharges become increasingly complex, it is unlikely that toxicity assessment based on individual constituents will be meaningful.

We are interested in the development of microbial toxicity tests such as that involving adenosine triphosphate (ATP) changes in natural populations of activated sludge micro-organisms involved in the degradation of waste substances. The tests must be simple, fast, cost effective, accurate and reproducible. An ATP bioassay using firefly luciferin-luciferase bioluminescence has been developed in our laboratory. The method detects changes in the universal energy compound of living cells when toxic substances are added to activated sludge mixed liquor samples. The technique has been found suitable for the screening of toxic substances such as heavy metals and we have been able to infer concentrations of particular toxic metals (e.g. chromium) in specific waste streams based on the toxicity bioassay. The ATP bioassay will be carried out in the near future in concert with two other methods, respirometry, which provides very useful information on the degree of biodegradation of the effluents at the wastewater works, and the Microtox^{TM} bioassay.

New Assays for Inhibition of Nitrification and Denitrification – A Comparative Study Applied to Industrial Wastewater

C. Grunditz, L. Gumaelius and G. Dalhammar

Dept. of Biochemistry and Biotechnology, Royal Institute of Technology, S-100 44 Stockholm, Sweden

Wastewater treatment plants in Sweden have requirements for nitrogen reduction and it is very important that the nitrogen removal processes are not disturbed. There is a great need for methods for investigating the inhibitory effects of industrial wastewaters on treatment plants. The best approach for assessing toxicity of complex wastes is to use biological methods. Pure culture organisms are preferred in order to yield reproducible results.

In this poster, three recently developed assays for the inhibition of nitrification and denitrification are presented. The test methods are based on three microbial reactions in the nitrogen removal process; ammonia oxidation, nitrite oxidation and nitrite reduction. Pure cultures of *Nitrosomonas, Nitrobacter* and a denitrifying strain *(No. 110)* are used. The assays are performed in 10 ml test tubes to which are added (1) a suspension of the bacterium in medium, (2) the water to be tested and (3) substrate in the form of ammonium or nitrite. The inhibition is calculated through comparison between the rate for a reference and for the water to be tested. (Grunditz and Dalhammar, manuscript; Gumaelius *et al.*, in press)

The three pure culture assays were compared in an investigation where the inhibition caused by 48 industrial wastewater samples was estimated. To investigate the correlation between the assays, a comparative study was performed using linear regression analyses. The correlation coefficients between each two methods were 0.63 - 0.73 which shows a rather poor relationship. This suggests that each reaction has its own inhibition pattern. Quantitative analyses of heavy metals (chromium, nickel, copper, zinc, lead and cadmium) in the wastewater samples were performed (SNV, 1994). Statistical analysis of the data was run. Zinc was shown to be a significant inhibitor in all the assays. The other significant metals were different for the three assays; copper for the ammonium oxidation, nickel for the nitrite oxidation and lead for the nitrite reduction. It is of interest that the methods respond differently on different metals. To sum up, the three assays give relevant, specific and reproducible results. All three methods are needed to give a complete picture of the inhibition of the different steps of nitrogen removal.

References

1. C. Grunditz, and G. Dalhammar, Assays for the inhibition of nitrification using pure cultures of *Nitrosomonas* and *Nitrobacter*. Manuscript.
2. L. Gumaelius, E. H. Smith, and G. Dalhammar, Potential biomarker for denitrification of wastewaters: Effects of process variables and cadmium toxicity. *Wat. Res.,* in press.
3. SNV (Swedish Environmental Protection Agency) (1994). Industribelastning på kommunala avloppsreningsverk. Med inriktning på nitrifikationshämning. Rapport 4376. In Swedish.

Ecotoxicological and Chemical Measurements of Municipal Wastewater Treatment Plant Effluents

J. Garric,[1] B. Vollat,[1] D. K. Nguyen,[2] M. Bray,[1] B. Migeon,[1] and A. Kosmala[1]

[1]Cemagref, Laboratoire d'écotoxicologie. 3 bis quai Chauveau, CP 220. 69336 Lyon Cedex 09. France.
[2]Société Lyonnaise des Eaux-Dumez, CIRSEE, 38 rue du Président Wilson, F-75231 Paris 05. France.

To assess the toxic impacts of three municipal wastewater effluents on the receiving waters we performed toxicity measurements and chemical analysis on 24 hours refrigerated composite samples of each effluent. These samples were collected between January 1993 and March 1995 from three wastewater treatment plants that receive wastewaters from domestic/commercial sources with different levels of contribution from industrial sources.

Conventional and specific chemicals analysis were performed simultaneously with acute (lethal *D. magna* test) and sub-chronic bioassays (reproduction *C. dubia* test and zebrafish embryo-larval stage test). We also measured a sublethal effect using the induction of the hepatic ethoxy-resorufin-O-deethylase enzymatic activity (EROD) of fish exposed to the effluents.

Chemical analysis, using GC-MS analysis allowed identification of a great number of organic compounds potentially toxic or inducing for the aquatic organisms. But, when they are known, the concentrations of each individual compounds never reached toxic levels. Nevertheless, we were able to detect toxic effect of the whole effluents using the battery of bioassays.

To attempt to characterise the main causes of the toxicity of the whole effluent, we used a toxicity identification approach. Additional experiments were carried out to measure the toxicity of various effluent fractions, after pH modifications or acid and basic/neutral extractions.

From chemical data and results of bioassays, the ammonia concentration appeared to explain a part of the lethal toxicity. But chronic toxicity tests carried out with whole effluents or extracted chemical fractions showed that organic pollutants, in spite of the low concentrations detected in the effluents, can also cause long-term effects. All the tested effluents induced fish EROD activity in the laboratory even though no PCBs and only low PAH concentrations were detected. Such a result also suggests that the use of this biomarker can help the field monitoring of these effluents.

Fish EROD Activity to Assess the Impact of a Wastewater Treatment Plant. Field and On-site Experimental Approaches

A. Kosmala, B. Migeon and J. Garric

Cemagref, Laboratory for Ecotoxicology, 3 bis quai Chauveau, 69336 Lyon Cedex 09, France

The effects of the effluent of a wastewater treatment plant (WWTP) were measured on the ethoxyresorufin-O-deethylase (EROD) activity of fish. EROD activity is included in the group of the P450-dependent mixed function mono-oxygenases. Effluents from WWTPs are known to contain some compounds such as PCB and PAH which induce EROD activity.

The WWTP under, study located in Châtillon-sur-Chalaronne near Lyon, receives domestic and industrial discharges and has a capacity for 6,000 inhabitants. Field and on-site laboratory experiments were carried out in June and September 1995.

Chub *(Leuciscus cephalus)* and stone loaches *(Nemacheilius barbatus)* were caught by electrofishing at three stations on the Chalaronne River (two stations upstream from the WWTP and one station downstream): ten males and ten females of each species. During the same period, juvenile rainbow trout *(Oncorhynchus mykiss)* and juvenile mirror carp *(Cyprinus carpio)* were placed in three basins supplied with the WWTP effluent and the Chalaronne water (a mixture of three different concentrations of effluent was made in the three basins: 0, 50, 100%. This experiment lasted 16 days: ten fish of each species were sacrificed at 4, 8 and 16 days and their livers were homogenized. Oxygen, temperature and ammonia were controlled.

EROD activity was measured on S9 supernatants of liver homogenates. Fish muscle and sediment from the river were analysed for PAH and PCB.

EROD activities increased significantly (in comparison with laboratory controls) for the carp exposed to the WWTP effluent, after four days exposure. The response is significantly linked to the concentration (50%, 100% of effluent). EROD activity is more relevant at four or eight days of exposure. Later, EROD activity seems to decrease. Carp seem more adapted to such monitoring because they are more resistant to temperature, hypoxia and ammonia encountered when exposed to effluent.

EROD response of fish from the river is rather complex. In June, there was an EROD activity decrease of female loach between the three stations and in September, on the contrary, the EROD activity of chub increased. Chub seems to be more inducible than loach. EROD activities of males and females are equivalent in September but not in June (probably linked to sexual maturation).

Influence of Organic Matter on the Response of Microtox™ Tests to Three Toxic Compounds

C. Ravelet, B. Vollat, J. Garric and B. Montuelle

Cemagref, Laboratory for Ecotoxicology, 3 bis quai Chauveau, 69336 Lyon Cedex 09, France

The Microtox™ test is often used for effluent monitoring applications as it is a rapid, reproducible and relatively inexpensive toxicity test. More and more attention is given to the part that organic matter could have in the apparent toxicity of toxic compounds in complex environments: sediments or wastewater for example.

We have tested the influence of three organic matters (OM) of various chemical nature (bacterial culture medium, sediment pore water, outlet of wastewater treatment plant) on the IC50 of various toxics: phenol, zinc and LAS. Increasing concentrations of OM (0, 30, 60 and 100%) have been used as preparative solution for the toxics. The 100% OM solutions were adjusted by dilution to 12 mg DOC per litre.

The IC50s of the three toxics mixed with organic matter were not significantly different from the reference IC50s of the pure toxics. Wastewater treatment plant effluent at high concentration (60 and 100%) increased the IC50 of LAS (70 mg l^{-1} instead of 30-50 mg l^{-1}). Conversely, the culture medium for bacteria decreased slightly the IC50 of $ZnSO_4$.

Therefore, it seems that organic matter does not have an important role in the expression of toxicity for the three toxics studied. However, the range of our tests do not allow to extrapolate and to generalise this lack of effect of organic matter: the chelation and the complexation capacity of organic compounds are variables, depending on their biochemical nature, the environmental parameters (pH, redox potential, contact time, etc.) and the toxic substance itself.

The Relationship between Measured Concentrations of Contaminants and the Toxicity of Leachates from a Contaminated Landfill Site

M. J. Mallett[1] and J. Sweeney[2]

[1]Euro Laboratories Ltd., 74, Sunderland Road, Sandy, Bedfordshire
SG19 1QY, UK
[2]Environment Agency Anglian Region, Harvey Street, Lincoln
LN1 1TF, UK

Groundwater pollution caused by leachate migration from former landfill sites near Helpston, Peterborough, has been the subject of comprehensive investigation by the National Rivers Authority (NRA EIR 1995). A major concern is the effect of leachates from the site on aquatic life in the receiving water environments. Euro Laboratories Limited have carried out work, under contract to the NRA Anglian Region, to determine the toxicity of samples of leachate taken from various locations (mainly boreholes) from the site on representatives of the aquatic community. These comprised algae (*Chlorella vulgaris*), an aquatic plant (*Lemna minor*), an invertebrate (*Daphnia magna*) and trout (*Oncorhynchus mykiss*).

C. vulgaris was the most sensitive to all samples tested except one, which was most toxic to trout. *D. magna* proved to be the least sensitive organism tested. *L. minor* was nearly as sensitive as *C. vulgaris* to most samples (except one). The *Lemna* test method had advantages over *C. vulgaris*. Most importantly it was not affected by the colour or turbidity of the sample.

In order to relate the toxicity seen to the identities and concentrations of chemical contaminants present, the NRA analysed for up to 79 chemical parameters in each of the samples. High concentrations of herbicides (mainly mecoprop) and ammonia were found. Toxicities of the determinands in the leachates, derived from literature values, have been added together in an attempt to account for the apparent toxicities of the samples.

There was a link between mecoprop concentration and toxicity to the plants, although in most cases toxicity was greater than expected from the measured concentrations of the herbicides. This discrepancy may have been due in large part to the inadequate toxicity data for the herbicides to *C. vulgaris* and particularly to *L. minor*. For trout and *D. magna* most of the toxicity seen could be attributed to the concentrations of ammonia present in the leachate samples. However, for rainbow trout one of the samples was considerably more toxic and for *D. magna* two were less toxic than expected. This may have been due to synergistic or antagonistic toxic effects among the components, and/or to the presence of unidentified toxicants.

Reference

NRA EIR 1995 Helpston Groundwater Investigations 1994/95. National Rivers Authority Anglian Region Environmental Impact Report. Report No. 2. September 1995.

Stonewort Cell: A Biosensor to Assess Wastewater Toxicity on Membrane and Whole Cell Levels

L. Manusadzianas[1] and R. Vitkus[2]

[1]Institute of Botany, Z. Ezeru 47, 2021 Vilnius, Lithuania
[2]Vilnius University, Dept. Biochem. & Biophys., Ciurlionio 21/27, 2009 Vilnius, Lithuania

Rapid, direct toxicity assessment of wastewater and chemicals is of great interest in pollution control. The toxicity based criteria tend not only to complement the chemical ones but to substitute them in certain cases. The biostressor-induced reactions on enzyme or membrane levels are generally fast, however, to match ecotoxicological requirements, they have to be correlated with the adverse effects on the whole cell or organism.

Stoneworts are widely distributed, grow abundantly in the ground plantations of clean fresh or brackish waters and are known as sensitive bioindicators. The giant charophyte cell separated from talloma and kept in artificial pond water maintains essential physiological characteristics for a long time. Due to its morphological features and easy-handling, the charophyte cell is a classic object in electrophysiology and cell biology, as well as being suitable for biotesting.

We have developed a rapid and quantitative biotesting method which is based on changes of electrophysiological response in *Nitellopsis obtusa* cells. Acute sub-lethal toxicity is evaluated by computer-assisted testing procedure that includes the registration of membrane parameters of up to 16 single algal cells simultaneously. Bio-electrical activity of living cells is measured according to K^+-anaesthesia method modified for multichannel recording with extracellular electrodes. The percentage decrease of averaged membrane potential during 0.5 - 2.0 h period was used as a parameter for the EC50 determination. This end-point of short-term membrane reaction was compared with the one of a long-term cell lethality response. The cell's death was judged according to the loss of turgor in semi-static tests of 96 and 192 h duration.

The data concerning rapid toxicity assessment of wastewater as well as of some reference substances will be presented. For the complex Vilnius wastewater samples the 0.5 h EC50s were approximately equal to 96 h LC50s. The respective values given for heavy metals showed higher sensitivity of lethality test.

Electrophysiological biotest has an advantage of being supplemented by the additional information on the alterations of other electrical membrane parameters. Some attempts in using a cell membrane conductance in biotesting will be discussed.

Development of Rapid Toxicity Assay for Paper *via* Microtox™ Test

A. Kahru,[1] L. Pollumaa,[1] I. Kulm[1] and E. Paart[2]

[1]Institute of Chemical Physics and Biophysics, Akadeemia tee 23, Tallinn EE0026, Estonia
[2]BIM KEMI AB, Box 3102 S-443 03 Stenkullen, Sweden

Photobacterium phosphoreum are naturally luminescent marine bacteria that generate light as a metabolic product. Any damage to the cellular metabolism caused by the toxicity of a substance can be monitored by measuring the change in light output from the bacteria, the degree of toxicity being proportional to the light loss. These bacteria (e.g., Microtox™ test bacteria) have been used for *in vitro* toxicity testing for more than 10 years, mainly for screening purposes of toxicity of industrial effluents.

A method that uses Microtox™ bacteria as test organisms has been worked out for the rapid analysis of relative toxicity of paper. Inhibition of light output of reconstituted photobacteria (INH%) after their exposure to paper extracts was used as toxicity endpoint. The efficiency of extraction of toxic substances (i.e., resulting INH% values) and reproducibility of the assay (% CV values) were used as evaluation criteria for different assay conditions: disintegration methods of paper, concentrations of disintegrated paper during extraction (1-4%), extractants (water, methanol, DMSO), extraction times (1-72 h) and exposure times (5-30 min) of resulting extracts to Microtox™ bacteria. The most appropriate assay scheme was worked out and used for the testing of relative toxicity of nine different types of paper.

Optical Fibre Toxicity Bio-sensor

D. F. Merchant,[1] P. J. Scully,[1] R. Edwards[1] and J. Grabowski[2]

[1]Division of Engineering and Science, Liverpool John Moores University, Byrom Street, Liverpool L3 3AF, UK
[2]Institute of Physics, Poznan Technical University, ul. Piotrowo 3, 60-965 Poznan, Poland

An optical fibre toxicity bio-sensor is presented for point or on-line monitoring of toxic effluents. The rate of hydrolysis of fluorescein diacetate by living micro-organisms is proportional to their metabolic rate, producing fluorescein. When excited with light with wavelength below wavelength 490 nm, this product fluoresces at 512 nm.

The rate of increase of fluorescent emission with time is related to the metabolic rate of the organisms which is affected by toxic substances in the cell environment. This effect has been experimentally verified for several heavy metals and work is under way to validate the method for other metals and organic contaminants.

Excitation light is generated by an electroluminescent film emitting in the range 450 to 505 nm, wrapped cylindrically round a glass tube containing an aqueous suspension of cells. Fluorescence is monitored over the sensitised surface of a 1 mm diameter plastic optical fibre immersed in the glass tube. A system of physical separation of the excitation and detection regions of the sensor removes the need for expensive optical filtering and increases the sensitivity of the detection assembly.

The fluorescent light entering the fibre is measured using a simple photo detector. The rate of increase of the light signal can be measured in minutes. The presence of toxins affects the time dependence of the light intensity.

The sensor gives a rapid early warning of the presence of toxins, and does not rely on a knowledge of the nature of the contamination. It is, therefore, particularly applicable to waste processing and effluent monitoring where the composition of the contaminants or their toxicity effects are unknown.

This sensor is rapid, compact, cheap, and rugged, as well as possessing all the advantages of optical fibre sensors such as resistance to electromagnetic interference. It can be configured as an on-line or a point sensor in a wastewater process, or as a portable, stand-alone sensor for sampling over minutes rather than hours. The technique is species independent and applicable to any biocenosis.

The Influence of Sewage Treatment Processes on the Oestrogenicity of Chemicals to Fish

G. Panter[1], R. Thompson[1] and J. Sumpter[2]

[1]Brixham Environmental Laboratory, Freshwater Quarry, Brixham, Devon TQ5 8BA, UK
[2]Dept. of Biology and Biochemistry, Brunel University, Uxbridge, Middlesex UB8 3PH, UK

A delicate hormonal balance controls reproductive cycles and secondary sexual characteristics in all vertebrates. In amphibia, birds, fish and reptiles, sex determination is under endocrine control, but is genetically established in mammals. This hormonal balance can be disrupted by environmental contaminants, affecting the reproductive capacity of the species concerned. Contaminants of greatest interest are those with oestrogen-mimetic activities, which can exert a feminising effect on target vertebrates.

In the aquatic environment, it has been shown that a large number of treated sewage effluents are oestrogenic to fish. Reproductive abnormalities and changes in vitellogenin (VTG) plasma levels (egg yolk precursor) have been demonstrated in rainbow trout (*Oncorhynchus mykiss*) and carp (*Cyprinus carpio*). Out of a probable 60,000 organic micropollutants, both natural and anthropogenic, present in the influent, it is uncertain which are responsible for the oestrogenic activity seen in the effluent. Furthermore, their activity may also be changed by the sewage treatment process. This project was designed to help bridge this gap in our understanding.

The biological treatment process of a sewage treatment plant has been simulated in the laboratory, using inocula from a local sewage works maintained on a synthetic sewage free from unknown chemicals. Selected oestrogenic chemicals were dosed into this simulated process to determine whether the microbial activity enhanced or diminished the oestrogenicty of the chemical. Effluent was supplied direct to fish exposure systems. Oestrogenicity was determined by comparing plasma VTG levels in fish exposed to the effluent with controls and those exposed directly to the chemical. VTG levels are low or undetectable in male and immature fish; any increase in plasma levels of VTG in these fish signified exposure to oestrogenic material.

Development of the carp VTG radioimmunoassay allowed plasma levels of VTG to be measured in the fathead minnow (*Pimephales promelas*). Plasma from the sheepshead minnow (*Cyprinodon variegatus*) did not cross-react with this assay. From these results the fathead minnow could be used as an alternative test species to the rainbow trout.

The biological process of a sewage treatment works and an artificial sewage were developed in the laboratory that supported life typically found in the biological treatment process. The resulting effluent, when exposed to fish via a semi-static or flow-through system was shown to contain no or little oestrogenic activity.

Preliminary work with a metabolic conjugate of natural oestradiol, oestradiol-3 glucuronide, indicated that this compound is inactive in direct fish exposure systems. However, after passing through the simulated sewage process this compound showed some oestrogenic activity. Further studies will be conducted on oestradiol-3 glucuronide to determine whether this was a true result, or due to rapid de-conjugation in the fish exposure system.

Evaluation of Two Enhanced Chemiluminescence Test Kits for Water Quality and Toxicity Testing

A. Colley, K. Wadhia and K. C. Thompson

Lab Services, Yorkshire Environmental, Templeborough House, Mill Close, Rotherham S60 1BZ, UK

Two commercial enhanced chemiluminescence (EC) test kits using the same technology were evaluated over a period of several months. The methods are based on monitoring the steady rate of chemiluminescence as luminol reacts with an oxidant in the presence of an enhancer. The horseradish peroxidase enzyme (HRP) acts as a catalyst in this method. The presence of a number of substances can result in a decrease in the chemiluminescent intensity.

The systems were found to be sensitive to certain toxins (e.g. phenol), but also very insensitive for some other toxins (e.g. chromium (VI) and aniline). In addition they can also respond to low concentrations of a few non-toxic substances (e.g. manganese and humic acid). The systems are also sensitive to antioxidants such as ascorbic acid.

Suitable quality control protocols were found to be essential in obtaining reliable results. It is **strongly** recommended that blank measurements using a reference water are carried out between each sample measurement. It is also recommended that duplicate measurements of each sample should be made and the mean taken. This protocol will help to highlight erroneous results (i.e. poor agreement) caused by operator and methodology errors. This can include drift due to changes in reagent activity, errors in sample preparation, pipetting reagent errors and operator bias errors. A standard toxicant should be run with each batch of samples.

The methods are considered to be more suited to use in buildings, rather than in true field situations especially in adverse weather conditions. The results were not significantly affected by sample pH (3 - 10) or temperature (12 - 25°C).

Some industrial effluents were tested using EC and conventional assays (Microtox, nitrification and respiration inhibition. The EC test method was found to be 2 - 200 times more sensitive than these other bioassay methods.

The EC test method could prove a useful and rapid screen prior to more time-consuming and costly acute bioassays.

For field use, the EC test kits are considered gross comparative tests for water quality. They are also suitable as a screening test for the rapid detection of pollutants in industrial processes or natural waters, (either up/downstream of a particular sample point, or periodic monitoring to observe change in quality). They have potential for detecting and rapidly tracing gross changes in water quality.

It was concluded that the technique has considerable potential, but the results needed careful interpretation.

The Fungicide Busan 30 WB Causes Lipid Peroxidation in Mussels Exposed to Leather Tannery Effluent

A. R. Walsh, P. Byrne and J. O'Halloran

Dept. Zoology, University College Cork, Ireland

The Ecotoxicological implications of leather tannery discharges have been poorly investigated. Of those studies that have been completed, much of the focus had centred on chromium toxicity, salinity, and oxygen depletion due to the inherent high BOD of the effluent. In comparison, the potential toxicity of fungicide discharges has not been addressed to any significant degree. Therefore, the degree of lipofuscin accumulation in the digestive cells of mussels in an estuary receiving untreated effluent from a tannery was assessed histochemically as a biomarker of oxidative stress. In addition, the concentrations of the peroxidative product malondialdehyde (MDA) was determined in cytosolic extracts. In the laboratory, mussels were exposed to the process fungicide Busan 30WB containing the active ingredient 2-(thiocyanomethylthio) benzothiazole (TCMTB) at concentrations of 50 and 100 μg l^{-1}. Groups of mussels were also exposed to chromium complexes which were also considered to be potential peroxidative agents, i.e. Cr(VI) and Cr(III)-albumen.

In the estuary, mussels were found to have elevated concentrations of lipofuscin in their digestive cells, with mussels closest to the outfall exhibiting the most intense histochemical staining. MDA concentrations were also elevated compared to the reference animals. In the laboratory, both chromium complexes gave negative peroxidative responses. However, exposure to TCMTB resulted in enhanced peroxidation in digestive cells at both exposure concentrations. In addition, ameobocyte proliferation was evident in the gills.

These results indicate that benzothiazole fungicides may represent a significant ecotoxicological threat in waters receiving leather waste.

Monitoring Compliance with a Toxicity-based Consent – The Case for Limit Tests

I. Johnson and P. Whitehouse

WRc, Henley Road, Medmenham, Marlow, Bucks. SL7 2HD, UK

Under the UK's Toxicity-based Consents programme for controlling complex effluent discharges to surface water, compliance with such consents will be judged on the basis of regular toxicity testing. Single concentration (limit) tests provide a cost-effective means of generating the required data in which a significant difference between the control and treatment groups specified in the consent would indicate a breach of consent. However, that decision needs to take account of the variability of the test method if false positive or false negative results are to be minimised which can result in either adverse commercial implications or environmental impacts which go undetected.

The poster shows how such errors can be minimised when limit tests are used for compliance monitoring. These involve (a) selecting an appropriate response level when the consent is derived, and (b) the use of a modified t test which includes a measure of within-laboratory variability - the main source of variability in compliance monitoring and the derivation of the 'reliable toxicity detection limit' (RTDL).

When such steps are taken, we suggest that simple limit tests offer a cost-effective approach to routine compliance monitoring with Toxicity-based Consents.

The Application of QSARs to Predict Environmental Hazard of Chemicals

A. B. A. Boxall

WRc, Henley Road, Medmenham, Marlow, Bucks. SL7 2HD, UK

The likely environmental hazard of chemicals can be assessed using scores which are comprised of measures of toxicity, sorption, solubility, volatility and biodegradability. These scores are usually calculated from experimental data, when no data are available default values are used. As the fate and effects of a compound in the environment are primarily dependent on its physical and chemical properties, one alternative to default values is to predict data from chemical properties using quantitative structure activity relationships (QSARs). The objective of this study was to determine the suitability of QSAR generated data for assessing the environmental hazard of chemicals.

Experimental data for toxicity, solubility, volatility and sorption of 47 compounds were obtained from the literature. This information was also predicted using QSARs. Experimental and QSAR-derived data were converted to hazard scores and compared. In general, the QSAR-based hazard scores were the same or very similar to scores based on experimental data, confirming the value of QSARs for predicting hazard. As the QSAR predictions were based entirely on chemical structure the approach has a number of applications, these include 1) pre-synthesis assessment of hazard; 2) prioritisation of substances for development and 3) selection of process chemicals.

Ecogeochemical Consequences of Contamination of the Nura River in Central Kazakhstan by Mercury-containing Wastewater from Acetaldehyde Production

M. Ilyuschchenko, S. Bulatkulov and S. Heaven

Kazakh State Academy of Architecture and Construction, Deptartment of Environmental Technology, 28 Ryskulbekov (Obruchev) St, 480043 Almaty, Republic of Kazakhstan

This paper presents a case study of a major example of industrial pollution in the former Soviet Union, and focuses on the problems of assessing the scale and impact of contamination. The Nura River (344 km in length) is the main waterway of the Tengiz-Kurgaldzhinskaya depression in Kazakhstan, and flows through the Karaganda industrial region and the city of Temirtau. The main industrial enterprises of Temirtau are the metallurgical factory, AO Carbide, and the thermo-electric power station KarGRES-1

The chief source of mercury contamination in the Nura is AO Carbide which produces acetyldehyde using a mercury catalyst. It began work in 1951 using German equipment received as reparations after World War II. Before 1976 there was no local treatment of the wastewater, and the mercury content reached 50 mg l^{-1} with an average annual discharge of 20 tonnes. In the 1950-1960's there were also discharges of untreated water from KarGRES-1 with up to 2.5 g l^{-1} of suspended solids. Nowadays the wastewater (mercury content 3-10 mg l^{-1}) is neutralised, treated with sulphide, settled and then mixed with other discharges from AO Carbide. As a result, the mercury content is reduced to 0.001-0.002 mg l^{-1}. According to average annual readings in the early 1990's, the mercury content in the Nura's water 0.5 km downstream from the discharge point was 0.001 mg l^{-1}, whereas data from independent observations indicated 0.02 mg l^{-1}.

A new type of alluvial sediment has formed in the river bed - technogenic mercury-containing silts. The total volume of silts for a distance of 100 km is 2.5 million m^3, of which 80 % is within 25 km of the discharge point. The composition of the silts does not change, except for the mercury content which reaches a maximum of 3 g kg^{-1}. The mass of mercury in the river bed is estimated at 140–160 tonnes, in a variety of mobile forms. During Spring floods, when the discharge is equal to 80% of annual flow, large quantities of silt are transferred not only along the river bed but also along the floodlands. The silts are a source of secondary contamination of the water. Mercury content decreases sharply in the first 15 km from the discharge point, but then gradually increases. The proportion of dissolved forms of mercury also increases in areas of intense biogeochemical activity.

Subject Index

Acartia tonsa, 46
Acetaldehyde, 287
Activated silica, 241
Activated sludge, 55, 75, 88, 241, 265
   respiration inhibition test, 78
   treatment, 219
Adsorption, 162, 169
   adsorption onto sludge solids, 160
Aeration, 182
Algentest IWF fluorometer, 256
Allyl thiourea (ATU), 58
Ammonia, 59, 158, 182, 236, 239, 274
   monitor, 56
   oxidation, 267
Amperometry mediated, 76
AMTOX, 54, 56
Aniline, 282
Anisus vortex, 146
Anodic Stripping Voltammetry, 168
API separation, 219
Aqua-tox Control
   Daphnia monitor, 256
   fish monitor, 256
Aquanox, 86, 256
Arbacia punctulata, 180
Arsenic, 236
Ascorbic acid, 282
Asellus aquaticus, 193
ATP bioassay, 265
Available dilution, 38
Aviation fuel, 147

Bacterial biofilm, 256
Benthic communities, 252
Benzo(a)anthracene, 196
Benzo(a)phenanthrene, 194
Benzo(a)pyrene, 196
Benzofluoranthene, 196

Benzothiazoles, 284
Best Available Techniques (BAT), 7, 149, 155, 176
Best Available Techniques Not Entailing Excessive Cost (BATNEEC), 34, 210
Best Practicable Environmental Option (BPEO), 26, 33, 155, 253
Best Practicable Technology (BPT), 176
Bioaccumulation, 9, 10, 151, 160, 192, 193
Bioavailability, 106, 258
Biochemical Oxygen Demand (BOD), 158
Bioconcentration factor (BCF), 151
Biodegradation, 160
   anaerobic, 162
Bioelectrial activity, 276
Biogeochemical activity, 288
Biological assessment, 104, 105, 120, 175, 178, 253
Biological survey, 27, 125, 252
Bioluminescence (*see also* Microtox & Lux), 34, 255, 261, 265
Biomarker, 20, 33, 107, 110, 111, 121, 140, 258
Biomonitoring (*see* Biological assessment)
Biosens-algae toximeter, 256
Biosensor, 74, 75, 85, 254, 276, 279
Biox 1000T, 256
Body burden, 140
BPMO (Benzopyrene mono-oxygenase), 111
Brachiodanio rerio, 269
Brook trout (*see Salvelinus fontinalis*)
Brown shrimp (*see Crangon crangon*)
Busan 30WB, 284

Cadmium, 132, 167, 169, 236, 255, 256
Carbaryl (Sevin), 141
Carbon Preference Index (CPI), 191

Carcinogenicity, 188
Cation exchange resin, 241
Cationic polymers, 158, 161
Ceriodaphnia dubia, 179, 180, 183, 239, 269
Certification programme, 37
Champia parvula, 180
Charophyte, 276
Chelation, 182, 214
Chemical Oxygen Demand (COD), 59
Chemiluminescence, 34, 255
  enhanced test, 85, 282
  reaction, 96
Chironomus, 124
  C. riparius, 34
Chlamydomonas reinhardii, 256
Chlorella vulgaris, 274
Chlorine dioxide, 162
Cholinesterase
  activity, 111
  inhibitors, 257
Chromium, 132, 171, 236, 282, 284
CHRONITOX, 34
Chrysene, 196
Chub (*see Leuciscus cephalus*)
Clean Water Act (1987), 176
Clearance rate, 129
Colour
  pollution, 158
  removal, 58, 60, 157, 158, 160
Cometabolism, 202
Compliance monitoring, 36, 39, 45, 126, 285
Consents (*see also* Direct Toxicity Assessment & Toxicity Based Consents), 26
  enforcement, 39, 176, 177
Copper, 132, 214, 236, 255, 256, 267
Copper sulphate, 90
Corophium volutator, 34
Crangon crangon, 232
Crassostrea gigas, 40, 46, 210
Cyprinodon variegatus, 180, 218, 281
Cyprinus carpio, 271, 280

Dangerous Substances Directive, 155
Daphnia contact test, 206
Daphnia life cycle, 206
Daphnia sp., 30, 34, 124, 239

D. magna, 40, 46, 78, 85, 153, 179, 204, 239, 256, 269, 274
D. obtusa, 204
D. pulex, 179
Denitrification, 267
DF-algae test, 256
Di(2-ethylhexyl) phthalate, 141
Dibutyltin, 141
3,4-Dichloroaniline (3,4-DCA), 47, 90
Dichloromethane (DCM), 215
Dichlorophenol, 255, 256
2,4-Dichlorophenol, 141
3,5-Dichlorophenol (3,5-DCP), 68, 76, 86
Dichlorvos, 141
Diesel fuel, 147
Diffuse sources, 139, 259
Dilution factor, 263
4,4-Dimethyl-1,3-dioxane (DN), 147
Direct Toxicity Assessment (DTA), 26, 36, 44, 45, 117, 122, 155, 209, 252, 276
Dissolved air flotation, 161
Dissolved organic matter (DOM), 168
Dredgings - remediation of contamination (*see also* Treatment), 264
Dye
  acid dye, 159, 161
  azo dye, 159, 162
  basic dye, 159, 160, 161
  direct dye, 159, 161
  disperse dye, 159, 161
  metal-complex dye, 159
  mordant dye, 159
  reactive dye, 159, 161
  residues, 158
  toxicity, 160
  vat dye, 159
  waste, 157, 158

ECLOX, 30, 34, 39, 40, 86, 94, 256
Ecological assessment, 105, 111, 156
Ecological relevance, 30
Economics, 10, 19, 217
Ecosystem health, 109
Effluent (*see also* Sewage)
  agricultural run-off, 118
  characterisation of toxicity, 38, 41
  chemical assessment, 177, 253, 261, 269
  consents (*see also* Direct Toxicity

Index 291

Assessment & Toxicity Based Consents)
  domestic, 2, 67, 98, 139, 209, 269
  food processing, 98
  industrial, 2, 67, 74, 84, 125, 139, 208, 217, 230, 265, 267
  leather tannery, 284
  oil-refining, 146
  petrochemical, 146
  pH, 237
  pH adjustment, 212
  produced water
    biodegradability, 263
    composition, 263
    toxicity, 263
  quality control, 200
  segregating of streams, 217, 229
  synthetic fibres, 238
  textile, 98
  toxic impacts, 1
  treatability study, 217, 239, 248
  treatment, 55, 201, 217
  treatment acclimatisation, 202
  treatment plant, 271
  whisky distillery, 255
Endrin, 141
Energy budget, 131
Enhanced chemiluminescence tests, 85
Enteromorpha intestinalis, 142
Entosiphon sulcatum, 239
Environment Act, (1995), 5
Environmental Assessment, 26, 253
Environmental Impact Assessment, 252
Environmental Protection Act, (1990), 5, 27
Environmental Quality Objective (EQO), 6, 149
Environmental Quality Standard (EQS), 26, 27, 149, 209, 230, 252
Escherichia coli, 75, 256
Ethoxy-resorufin-O-deethylase (EROD), 111, 269, 271
Ethylenediamine (EDA), 239
Ethylenediaminetetra-acetic acid (EDTA) (*see also* chelation), 182, 213
Eu-cyano bacterial electrode, 256

Fathead minnow (*see Pimephales promelas*)

Federal Pollution Control Act, (1972), 178
Feeding rate, 120, 126, 129
Fish monitor, 256
Fixed limit approach, 6
Fluoranthene, 194
Fluorene, 196
Fluorescein diacetate, 279
Fluorescence, 143, 255, 279
Fluoride, 236
FluOX test, 256
Fucus vesiculosus, 143
Fugacity, 15
Fulvic acids, 171
Fungicide
  benzothiazole, 284
  Busan 30WB, 284

β-Galactosidase, 257
Gammarus pulex, 34, 118, 193
Gasworks site, 260
General Quality Assessment (GQA) Scheme, 155, 252
Golden Orfe (*see Leuciscus idus*)
Granular Activated Carbon (GAC), 217, 239, 241
Ground water, 261
  toxicity, 118, 260
Guppy (*see Poecilia reticulata*)

Hazard scores, 286
Horse radish peroxidase (HRP), 85, 96, 96, 256, 282
Humic acid, 282
Humic compounds, 168
Hydrocarbons, 140, 141, 188, 193
  aliphatic, 188
  polyaromatic (PAH), 188, 189, 271

Immobilised cultures, 55
Inhibition, 201
  load, 201
  of acetylcholinesterase, 122, 140
  of activated sludge respiration, 88
  of bacterial respiration, 201
  of denitrification, 267
  of nitrification, 54, 55, 65, 69, 86, 87, 88, 158, 165, 201, 206, 267, 282
  of photosynthesis, 143
  of treatment, 201

In situ tests, 112, 116, 125, 189, 192, 194, 237, 258
Integrated Pollution Control (IPC), 2, 26, 27, 33, 155, 231, 252
Integrated Pollution Prevention and Control (IPPC), 155
International Commision for the Exploration of the Seas (ICES), 258
International Standards Organisation (ISO), 258
Ion leakage, 142
Irgarol, 143

Kjeldahl nitrogen, total (TKN), 242, 244

Lac-z receptor gene, 257
Leachate, 274
migration, 274
Lead, 132, 167, 169, 236, 267
Lemna sp., 206
 L. minor, 85, 274
Leuciscus cephalus, 271
 L. idus, 256
Lindane, 141
Linear alkyl benzenesulphonate (LAS), 273
Lipofuscin, 284
Lux, 261
Lux-modified cultures, 255
Lymnaea peregra, 189, 193
Lysosomal integrity, 111

Macoma balthica, 192
Malathion, 122
Malondialdehyde (MDA), 284
Manganese, 171, 282
Maximum Tolerable Concentration, 150
Mecoprop, 274
Menidia beryllina, 180
 M. menidia, 180
 M. peninsulae, 180
Mercury, 236, 287
MERIT Respirometer, 64, 87, 89
Metallothionein, 111
Metals, 132, 167, 182, 183, 214, 236, 261, 264, 265, 267, 277, 279
 adsorption of, 169
 complexation, 167, 169, 170
Methyl-naphthalene, 194
Microbial toxicity tests, 265

Microcystis aeruginosa, 239
Microtox, 30, 39, 40, 46, 77, 85, 86, 95, 123, 210, 221, 231, 255, 256, 260, 261, 263, 273, 278, 282
 influence of organic matter, 273
 Microtox-os, 256
 sediment testing, 237
Mirror carp (*see Cyprinus carpio*)
Mixed function oxidase, 111, 271
Mixed Liquor Suspended Solids (MLSS), 56
Mixing zone, 123, 134
Mossel monitor, 256
Mussel (*see Mytilus edulis*)
Mussel Watch, 18, 194
Mutagenicity, 188
Mysidopsis bahia, 180, 218, 224
Mysid (*see Mysidopsis bahia*)
Mytilus edulis, 34, 125, 139, 192, 284

Narcosis, 135, 136, 141, 257
National Pollutant Discharge Elimination System (NPDES), 176, 218
Nemacheilius barbatus, 271
Neurotoxin, 141
Neutral red assay, 142
Nickel, 132, 236, 267
Nitellopsis obtusa, 276
Nitrate, 59, 164
Nitrification, 55
 inhibition, 54, 55, 65, 69, 86, 87, 88, 158, 165, 201, 206, 267, 282
Nitrifying bacteria, 55, 88, 202
Nitrite, 59
 oxidation, 267
 reduction, 267
Nitrobacter, 267
Nitrosomonas, 267
Nitrotox, 66

Octanol-water partition coefficient, 12, 141
Oestradiol, 281
Oestradiol-3 glucuronide, 281
Oestrogenicity, 21, 254, 280
Oil
 multigrade oil, 147
 production platform, 263
On-line monitoring, 54, 55, 59, 74, 203, 254, 279

Index

Oncorhynchus mykiss, 30, 40, 78, 180, 210, 256, 271, 274, 280
Organics
  non-polar, 182, 215, 219
  polar, 182
Organisation for Economic Cooperation and Development (OECD), 258
Organo-copper, 215
Oxygen
  consumption, 255
  debt, 129
  uptake, 65
  uptake rate, 55, 89

Pacific oyster (*see Crassostrea gigas*)
Pentachlorophenol, 141
Pesticide, 140
  organophosphorus, 122, 140
  pyrethroid, 120, 141
pH, 55, 164, 181
Phaeodactylum tricornutum, 40, 129
Phenanthrene, 194
Phenol, 90, 273, 282
Photobacterium phosphoreum, 85, 86, 206, 210, 255, 256, 262, 263, 278
Photosynthetic efficiency, 143
Physiological rate, 126, 128
Phytane, 192
Pimephales promelas, 179, 180, 280
Plaice (*see Pleuronectes platessa*)
Planorbis rubrum, 146
Pleuronectes platessa, 30
Poecilia reticulata, 153
Polyacrylamide polymer, 58, 60
Polyaromatic hydrocarbon (PAH), 188, 189, 271
Polychlorinated biphenyls (PCB), 9, 10, 271
POLYTOX, 95, 256
Polyvinyl alcohol (PVA) matrix, 55
Population dynamics, 107
Potassium hexacyanoferrate, 74
Precautionary Principle, 6, 10, 23
Predicted Environmental Concentration (PEC), 36, 38
Predicted No Effect Concentration (PNEC), 36, 38, 252
Priority pollutants, 182
Pristane, 192
Protocol guidelines, 42

Pseudomonas fluorescens, 255, 256, 261
P. putida, 75, 206, 239, 256
Pyrene, 194

Quality Control, 30, 45, 51
  Shewhart control system, 90
Quantitative Structure Activity Relationship (QSAR), 105, 136, 140, 286

Radioactive waste, 3
Rainbow trout (*see Oncorhynchus mykiss*)
Randox, 86, 95
Rapid toxicity testing (*see also* Bioluminescence & Chemiluminescence), 30, 36, 74
  algae, 206
  screen, 278, 282
Reducing agents, 182
Refractory toxicity assessment (RTA), 221
Regulatory control, 26, 27, 31
Reliable toxicity detection limit (RTDL), 285
Remediation, 260, 264
  techniques - physical, 264
Reservoir, 167
Residual Maximum Likelihood (REML), 48
Respiration
  inhibition, 65, 67, 86, 206, 282
  rate, 126, 130
Respiratory blockers, 257
Respiratory uncoupling, 140
Risk assessment, 36, 39
Risk management, 156
Rodtox, 95
Rolling tube, 206
Root elongation, 206

Safe Drinking Water Act, (1974), 176
Safety factors, 153
Salvelinus fontinalis, 180
Scenedesmus quadricauda, 239
Scope for Growth, 107, 127, 134, 139, 192, 203
Scophthalmus maximus, 30, 40, 210, 232
Screen-printed electrodes, 75
Sediment, 191
Selenastrum sp., 30
  S. capricornutum, 40, 153, 180

Self monitoring, 38
Sensitivity, 30, 38, 38, 45, 85, 86, 135, 136, 152, 203, 211, 234, 282
Sewage (*see also* Effluent & Treatment), 118
  outfall, 127
  sludge, 158, 261
  treatment, 96, 158, 280
  works performance, 84
Sheepshead minnow (*see Cyprinodon variegatus*)
Shell (bivalve)
  length, 130
  valve activity, 255
  volume, 130
Silverside (*see Menidia beryllina*)
Skeletonema sp., 30
S. costatum, 78, 210
Sodium thiosulphate, 182
Solid Phase Extraction (SPE), 182, 214
Sparging, 182
Standard Operating Procedures (SOP), 30, 258
Stiptox norm, 256
Stone loach (*see Nemacheilius barbatus*)
Stonewort, 276
Stormwater, 188
Surface water toxicity, 260
Synechococcus, 256

2-(Thiocyanomethylthio)benzothiazole (TCMTB), 284
Tisbe battagliai, 30, 34, 77, 210, 263
Toluene, 239
Total organic carbon (TOC), 263
Total oxidised nitrogen (TON), 163
Toxalarm, 256
TOXALERT, 256
Toxicity
  chronic sub-lethal, 33, 258
  endpoints, 180
    developement, 33
    juvenile growth, 33
    reproductive capability, 19, 33
    ventilation rate, 255
  limit, 36, 39, 45, 234
  modes of action, 139, 200
  narcotic mode of action, 135, 136, 141
  of paper extracts, 278
  sub-lethal effects, 33, 125, 138

surface water toxicity, 260
targets, 27, 38
test methods, 104, 179
  chronic, 180, 258
  control (reference), 205
  established tests, 36, 42
  in situ assays, 116, 125, 189, 192, 194, 237, 258
  limit tests, 285
  media, 47
  methods manual, 43
  microbial tests, 265
  multi-species tests, 259
  multiple test protocols, 200
  Nitrotox, 66
  oyster embryo test, 30, 232
  plant growth test, 204
  precision, 30, 44
  rapid tests, 30, 36, 42, 74, 85, 92, 94, 206, 261, 282
  repeatability, 30
  reproducibility, 30
  screening tests, 84, 94, 203
  selection of, 28, 37, 39, 203
  specificity, 204
  sub-lethal tests, 43, 258
  variability, 45, 48
  Zebrafish embryo-larval test, 269
toxicity unit (TU), 88, 227
Toxicity based consent, 117, 117, 155, 285
Toxicity based criteria, 26, 31
Toxicity Identification Evaluation (TIE), 177, 181, 208, 209, 211, 219, 230, 269
Toxicity Reduction Evaluation (TRE), 28, 36, 39, 101, 176, 181, 211, 217, 218, 219, 238, 243
Toxics control, 175
Toxiguard, 256
Treatment (*see also* Activated sludge), 55, 201, 217
  acclimatisation of, 202
  activated sludge, 219
  aerobic biological, 210, 241, 243
  air stripping, 241
  API separation, 219
  coagulation/flocculation/ sedimentation, 160, 162
  colour removal, 58, 60, 157, 158, 160
  dissolved air flotation, 161
  equalisation, 219

Index

filtration, 227
granular activated carbon (GAC), 217, 239, 241
macroreticular resin, 241
nutrients, 223
optimisation, 200, 217
oxidation, 162
ozone, 162
percolating filter, 161
porous pot, 206
reduction, 162
secondary clarification, 219
strategy, 218
tertiary, 98
treatability study, 217, 239, 248
ultrafiltration, 162
Tributyltin, 20, 140, 141
Trichloroethylene, 4
2,4,5-Trichlorophenol, 141
Trophic levels, 38, 40, 258
Turbot (*see Scophthalmus maximus*), 40

United States Environmental Protection Agency (USEPA), 175

Uronema parduczi, 239

Vibrio fischeri (*see also Photobacterium phosphoreum*), 255, 256
Vitellogenin (VTG), 280
Vitox, 161
Volatile organic carbon, 263

Wastewater (*see* Effluent)
Water mixing zone, 123
Water Quality Criteria, 177
Water Quality Objectives, 150
Water Quality Standards, 176
Water Resources Act, (1991), 27
Whole Effluent Toxicity (WET), 126, 175, 178

Yeast, 257

Zebrafish (*see Brachiodanio rerio*)
Zinc, 132, 167, 169, 236, 255, 256, 267, 273
Zinc sulphate, 47, 90
Zone of deterioration, 28